T0325563

Risk Management for Project Driven Organizations

A Strategic Guide to Portfolio, Program and PMO Success

Andy Jordan, PMP

Copyright © 2013 by Roffensian Consulting Inc.

ISBN-13: 978-1-60427-085-3

Printed and bound in the U.S.A. Printed on acid-free paper.

10 9 8 7 6 5 4 3 2 1

Library of Congress Cataloging-in-Publication Data

Jordan, Andy, 1971-
  Risk management for project driven organizations : a strategic guide to portfolio, program and pmo success / by Andy Jordan.
     pages cm
  Includes index.
  ISBN 978-1-60427-085-3 (hardcover : alk. paper) 1. Risk management. 2. Project management. I. Title.
  HD61.J66 2013
  658.15'5--dc23
                              2013010897

Phone: (954) 727-9333
Fax: (561) 892-0700
Web: www.jrosspub.com

Dedication

For Diane, who took a huge risk and made all of this possible.

Table of Contents

Preface

Running a business is a risky endeavor. It doesn't matter whether we are talking about the smallest start-up or the largest multinational, the decisions that a business makes put the success of the organization on the line every day. However, there's no alternative—it may be a cliché, but it's no less true that *nothing ventured, nothing gained*—if you don't take risks then you can't succeed.

However, that doesn't mean that a company should just accept those risks as part of what it takes to run a business. The risks need to be identified, categorized, analyzed, and managed. This is a well understood concept when it comes to the normal operations of a business—we put audits, management controls, checkpoints, and any number of additional safeguards in place to try and mitigate risks or identify when things are starting to go wrong.

The same is true in our projects. We undertake a series of formal processes to ensure that all of the risks that could impact the project are captured and appropriate strategies are developed to respond to those risks—reducing or eliminating the likelihood of them occurring, minimizing the impact on the project if they do occur, and developing contingency plans should they be needed.

This leaves the tactical elements of an organization fairly well covered—the operational and project functions that form the *sharp end* of what a company does, but it doesn't address the most significant areas of our organizations. Projects and operations as low-level activities are collectively the execution methods we choose to achieve the corporate goals. For commercial ventures, those goals are some combination of increased market share, reduced costs, increased revenue, or a similar measure that directly translates into corporate and financial growth.

It's much more effective to conduct risk management at the strategic levels where these goals are set and where the high-level planning occurs—specifically at the portfolio and program levels. The portfolio is the organizational vehicle of change—it's the umbrella structure that encompasses all of the initiatives the organization executes within a given time period to change the way it does business (anything that isn't a continuation of operations). A program extends that concept over a period for a subset of the portfolio, ensuring consistency in the way work is carried out in order to achieve a subset of the corporate goals.

Yet this doesn't happen. Routinely, organizations will set objectives based on the most rudimentary planning and in doing so expose themselves to levels of risk that would horrify them if set out in black and white.

That's why I wrote this book. I wanted to present a model that begins by understanding and quantifying the organization's risk environment and compares that to the concept of the organization's risk capacity. Then I wanted to move on to helping the organization proactively determine its risk tolerance and the relative importance of risk against all of the other factors that define how the organization's portfolio execution operates. Only when that's complete can an organizational risk management model be defined and implemented with any hopes of success.

This book assumes that the reader has an understanding of risk management, but it is not written solely or specifically for project and program managers. They will find some familiar concepts here, and there is advice on how to structure and execute an organizational risk management project, but those projects can only be successful with organizational buy-in. I hope that this book will be read by the executives who make the decisions about which initiatives your organizations will execute—they are significantly impacting the risk exposure whether they are conscious of it or not. I hope that this book will be read by portfolio management and project/program management office (PMO) resources—they control the execution of an organization's strategy. If risk management is not front and center in their minds at more than just the lowest level, then they aren't managing risks as effectively as possible. Finally, I hope that this book will be read by stakeholders—the decisions that they make will have an impact on the overall risk exposure, and a more complete awareness will lead to better decisions.

Of course, risk management is only one element of portfolio execution, and I am not going to claim that a well-constructed and well implemented organizational risk management approach is going to solve all of the challenges that an organization faces. An organization should consider an overall organizational-level portfolio execution methodology that applies similar concepts to all elements of the way that their strategies are executed. This will allow organizational risk management to become an integral part of an enterprise-wide, portfolio-focused approach to the manner that organizations drive change and growth, and that should be the ultimate goal.

I would love to hear your experiences with organizational risk management. You can contact me at andy.jordan@roffensian.com to let me know how the concepts are helping your organization to improve its strategy execution.

Acknowledgments

Writing a book can be a solitary endeavor, especially one on this topic as not all my friends share my passion for organizational risk management! However this book only exists because of the help, support, and assistance of a number of people, all of whom deserve much credit.

First, to Cam McGaughy, who first gave me a voice and the forum for that voice to be heard—without his support and belief in my early writings this book would never have happened. Next to Dave Garrett, who gave me the opportunity to be a part of his incredible *Project Pain Reliever* book—an honor that ultimately led to this point and for which I will always be grateful. I would also like to acknowledge my publisher, Drew Gierman, whose belief in me is the reason why this book has come to life.

When it came to the specifics of this book, there are a number of people who have made it what it is. Natalie Simms produced some great graphics that took my ugly process flows and made them coherent, understandable, and yes, Natalie, *pretty*. Kim Caughlin came up with the right words at the right time (and probably doesn't even realize it) to keep me on track and focused. Pauline Drywood, the best sister a brother could have, helped me to realize what this book meant.

I am also extremely grateful to all of my readers on ProjectManagement .com and ProjectsAtWork.com for their ongoing support and thought-provoking commentary and discussion. The thoughts and ideas in this book have been shaped in part by each of you.

The biggest thanks has to go to my wife who not only tolerated the long evenings on the laptop, she also edited this book while project managing (and risk managing) a major kitchen remodel that seemed immune to all attempts at schedule management—no man has the right to be as lucky as I am.

And thanks to you for reading this book; it is truly appreciated.

I thank no one for the errors and omissions—they are all my own work.

About the Author

Andy Jordan, PMP

Andy Jordan is president of Roffensian Consulting, an Ontario, Canada based management consulting firm with a strong emphasis on organizational transformation, portfolio management, and PMOs. Andy has assisted organizations in all aspects of portfolio, program, and project execution, as well as PMO structure and process. He has a track record of success managing business-critical projects, programs, and portfolios in Europe and North America in industries as diverse as investment banking, software development, call centers, telecommunications, and corporate education. He has also provided guidance on corporate strategy and has acted in operational c-level roles.

This seasoned business professional started his career in retail and private banking in the United Kingdom where he was awarded an associateship in the Chartered Institute of Bankers. His project management career began by accident when he was asked to take over an initiative partway through, and he discovered both a love and aptitude for the discipline.

Mr. Jordan earned his certification as a project management professional (PMP) from the Project Management Institute and continued managing projects and programs after immigrating to Canada. He developed an impressive résumé managing business-critical initiatives in various industries. Andy then moved into portfolio management and PMO leadership and rapidly developed a reputation for turning around troubled project-execution functions and delivering meaningful business results while maintaining and developing team performance and morale. He launched and relaunched a number of PMOs and project execution approaches before founding Roffensian Consulting.

Andy Jordan is a well-known author on project management and related topics. His literary works have been printed in industry and corporate publications on six continents. Andy is a prolific writer with new articles appearing weekly on ProjectManagement.com (formerly gantthead.com) with an audience of nearly 600,000 IT project managers and executives, and for ProjectsAtWork.com with an audience of more than 120,000 program and portfolio managers. He was also a contributing author on the widely praised book *Project Pain Reliever: A Just-In-Time Handbook for Anyone Managing Projects* edited by Dave Garrett.

Andy Jordan is a sought-after speaker and moderator for both in-person and web-delivered events for private clients and industry associations. He is an accomplished instructor on project management, risk management, leadership, and communication-related subjects. His hallmark is a pragmatic approach that recognizes that processes and methodologies have to exist and succeed within the real world of corporations and agencies.

Introduction

Trying to write a book on organizational risk management presents a number of challenges. Aside from the rather obvious challenge of writing a book on a topic that many people will find to be somewhat less than stimulating, there is the problem that we are dealing with terms meaning different things in different organizations. Project is a fairly well accepted term, and program isn't too far behind, but when we get to PMOs, it's a whole different story. There won't even be agreement on whether the acronym stands for project management office, program management office, or even portfolio management office. Try to get agreement on the scope of a PMO's reach or functions and life starts to get complex.

The term portfolio is no better. For some it is a snapshot of all of the projects underway within an organization, and for others it's the projects associated with a PMO. Now we're back to that challenge. Even risk management means different things to different organizations, with varying perceptions of where it starts and finishes—both in terms of a project lifecycle and in terms of scope of work.

I have tried not to get bogged down in these issues, which are ultimately nothing more than filters that you will apply to your interpretation and application of the concepts I set out here. Where I think that it's important to define terms, roles and responsibilities, or scope for the purposes of the book I have done so, everywhere else feel free to apply your own definitions. I've been more prescriptive when it comes to the framework for my organizational risk management process, but again, you will want to tailor and adjust this to your own unique needs.

I have broken this book into three distinct sections:

1. A section on organizational risk theories and concepts as well as the development of an organizational risk profile. This section provides a foundation for the rest of the book and helps to

provide context for the process itself—the best process in the world won't work if the environment is not ready for it.

2. A section that defines a detailed organizational risk management process framework and looks at some of the unique organizational risk management challenges that have to be considered at the portfolio and program levels, as well as in the PMO. This section also considers how project level risk management is impacted by a more strategic process.

3. A section on implementation that provides guidelines for understanding your organization's readiness for strategic risk management. It helps you to develop an implementation plan for taking the risk management framework from the previous section and evolve it into a specific process that works for your organization's needs, and then guides you through the process development and implementation. Some of the concepts here can be applied to process development projects in general. This is deliberate—you don't want to have a custom process development process for risk.

These sections are inextricably linked and are intended to build on one another as I guide you from concepts to a process model and then on to the application of that model in your unique environment. If your organization doesn't have a solid grasp on its risk environment, or if the idea of a constraints hierarchy and where risk fits in that hierarchy are not understood, then you are likely to face a number of frustrations; more significantly the potential for success will be limited. Before you try to implement an organization-wide strategic risk management approach, I strongly advise you to ensure that the foundation is in place. It's better to delay the implementation of the process until the organization is ready for the change than it is rush the process and suffer failure that damages the credibility of the concept.

By purchasing this book you are also entitled to access the Web Added Value™ Download Resource Center that is available at www.jrosspub.com. There you will find a number of examples of the templates and tools that are referred to in the book. They should help you to get a kick start on your implementation.

A Word on Risks

Before we get into the meat of the book, let's take a couple of minutes to look at risk at the most fundamental level. A risk can be defined as:

A potential event or occurrence, which, if it occurs, has a positive or negative impact on the project's objectives.

That's a pretty good starting point for us, and there are a couple of concepts there I would like to ensure you understand before we go any further.

First, I want to separate the concepts of risks being problematic and risks being bad. Risks are generally always problematic—they cause challenges we have to deal with. We either have to manage them as risks or we have to deal with the impact they cause if they become real; sometimes we have to do both. However, that does not mean that they are automatically bad—a risk could have a positive outcome for the project. A negative risk should more accurately be described as a threat, and a positive risk is an opportunity. Inevitably, the focus is on threats when we manage risks. This book has a similar focus but don't lose sight of positive risks—they can be just as significant. Think back in your career, or in your personal life, and you'll be able to identify a number of missed opportunities—those were positive risks that you failed to manage successfully.

Second, let's look at the concept that to be considered a risk an item has to impact a project's objectives. We'll be replacing project with some other words in this book, but the idea that there has to be a degree of significance to the risk is a good one. Often risk management activities set an arbitrary threshold level, and anything that falls below that threshold is ignored. The threshold or standard concept is a good one, but it shouldn't be arbitrary; it has to be based on a meaningful, real measure.

Set that bar too low and you will find yourself spending a significant amount of time managing risks that will have minimal impact on your initiative; set it too high and you will be blissfully unaware of the problem that is about to take you down. Tying the standard directly to the objectives makes things clear and simple to understand—if it can impact the outcome then we need to care about it!

We'll use these concepts throughout the book so make sure that you are comfortable with them before going further.

At J. Ross Publishing we are committed to providing today's professional with practical, hands-on tools that enhance the learning experience and give readers an opportunity to apply what they have learned. That is why we offer ancillary materials available for download on this book and all participating Web Added Value™ publications. These online resources may include interactive versions of material that appears in the book or supplemental templates, worksheets, models, plans, case studies, proposals, spreadsheets and assessment tools, among other things. Whenever you see the WAV™ symbol in any of our publications, it means bonus materials accompany the book and are available from the Web Added Value Download Resource Center at www.jrosspub.com.

Downloads for *Risk Management for Project Driven Organizations: A Strategic Guide to Portfolio, Program and PMO Success* include:

- *A Risk Management Plan Template*, with guidelines for a portfolio (or program) risk management plan format
- A Risk Summary Template, with guidelines for an individual portfolio (or program) risk to be detailed
- *A Contingency Plan Template*, with guidelines for a contingency plan including sections for recording actions on triggering
- *A Risk Identification Checklist* for people identifying risk candidates as part of the individual review step in risk identification
- *A Risk Analysis Checklist* for people conducting risk analysis or organizational risks
- *A Process Assessment Checklist* to help with the assessment of existing organizational processes as part of the project to develop and implement organizational risk management
- *A Process Development Plan Template* to assist with the development of processes and tracking of progress by team members

SECTION 1

Business Level Risk

You are an exceptional risk manager. Every day you make numerous decisions that require an analysis of the likelihood and impact of different possible results, and your actions are driven in part by the outcome of that analysis. Our education and training are geared around trying to make these analyses automatic in favor of the conservative option—not crossing the street unless there is a Walk sign, not accelerating at a yellow light, leaving home in time to ensure that we aren't late arriving at work, etc., but the decision is still ours to make. If we want to leave home late, accelerate through every light as it is changing, and then dodge traffic jaywalking between the parking lot and the office, we have the ability to make the decision to do so, with the understanding that the risk of a negative outcome is higher than if we were to leave home a few minutes earlier and take a more conservative approach.

Your environment will also impact the decisions that you make—it offers additional input into the risk analysis. For example, you will drive slower on a snowy night than you will on the same road on a sunny day. Finally, your motivations will impact your risk analysis—it's easier to resist the ice cream sundae when you are feeling energetic and positive than it is at the end of a bad day when nothing seemed to go right.

When you spend a few minutes thinking about it, there are literally hundreds of decisions a day that involve some degree of risk analysis, and yet few of those analyses are taken consciously. The risks are simply processed alongside everything else, and you either hit the brake pedal or the gas pedal when you see the light start to change, depending on the outcome of all those calculations. There is minimal, if any, conscious effort put into the calculations.

The same is true in organizations. Virtually every decision that the executives of an organization make will require some degree of risk analysis, but in most cases, it's not a formal process unless the decision is considered to be major. Instead, it's just *part of the job*, one of the many variables that go into the responsibilities of an executive. In fact, if we think back to the concepts we explored in the introduction, we said that to be considered a risk there had to be the potential to impact objectives, and even CEOs of Fortune 500 companies make their share of fairly innocuous decisions. There may be degrees of uncertainty associated with those less critical decisions, but if things don't go according to plan, the impact won't affect the company's ability to achieve its objectives.

External Risk Environment

How can an executive be sure whether their decisions are insignificant or potentially business destroying? They need to understand the risk environment within which they operate, just like you need to understand the risk environment within which you operate when you are driving that car and deciding what to do at the changing traffic signal.

For organizations, that environment consists of a number of variables outside its direct control but that still have the potential for dramatic impact. Some of these categories are related to the company's own internal risks, and some are completely independent. In most cases, there are opportunities to influence and control some of these external risk categories, but that's risk management and we're getting ahead of ourselves.

The major categories of external risks are shown in Table 1.1. You can see from that list that they collectively cover virtually everything around the company—its physical locations, its relationships with all external stakeholders, and its markets. That's not coincidental. Organizations don't exist in a vacuum, and the way that they interact with their environments will create new risks and influence existing ones.

In many cases these risks are fairly *slow moving*—changes to regulatory frameworks tend to be planned months or years ahead. Governments change generally only every few years, and even then tend to evolve rather than revolutionize; economic growth or contraction usually has warning signs ahead of the main impacts. This often results in a degree of organizational complacency when considering these risks. If there's no upcoming election then political risks get ignored. If the latest round of regulatory reporting improvements happened last year then the assumption is that they will be stable for the next couple of years at least.

Table 1.1   External risk categories and descriptions

Category	Details
Competitive	The competitive environment in which the company operates is an obvious area for external factors to impact. A new feature from a competitor, an aggressive pricing strategy, or a new entrant into the market can all affect the organization's risk environment. On the flip side a competitor's failed product can have an equally dramatic impact.
Economic	The economic situation can impact an organization's risk environment—the impact on sales and revenue, the ability to maintain investment and staffing levels, the speed with which they can recover, etc.
Geographic	An organization's location(s) contribute to their risk environment, with impacts varying from proximity to resources, communication links and markets to the number of competitors for resources. There can be an assumption that these factors are fairly stable, but that's not the case at all—earthquakes, tsunamis, hurricanes, etc., fall into this category as well and the potential for exposure contributes to the risk environment.
Investor	Financial risks are generally seen as internal to an organization, and obviously the company needs to take responsibility for its financial performance. However, there is a closely connected, but distinct, external risk—the concept of investor risk. This may be from one or more major shareholders, from a major external private company investor, or from a mass uprising from smaller shareholders.
Political	The political climate that an organization finds itself in can have a tremendous impact, a change from a right wing to a left wing government (or vice versa) can result in changes to everything from taxation to workers' rights and subsidies to export restrictions. In areas of political instability the impacts may be even more dramatic.
Regulatory	All industries are subjected to a degree of regulation, but for many the regulatory demands can introduce tremendous risk, especially if those demands change. Import/export regulations, reporting requirements, safety measures, etc., can all change the risk environment significantly. Public sector organizations can sometimes be just as heavily regulated with restrictions that do not necessarily follow logic. Regulatory risk directly connects to the internal risks around compliance.
Reputational	How the company is perceived has a major effect on the risk environment that it exists within. This category can cover everything from perceived quality of products and service to reputation as a corporate citizen. In the current reality of instant and global communications and the ability for any individual to have a voice through blogs, Twitter and the like, this is something that can have even greater impact than in the past. This is an external risk because the organization cannot control what is said about it.

Category	Details
Societal	One of the most obvious examples of societal risks today is the retiring baby boomers and the skills shortage that will result for many organizations. Other high profile elements of this category are the expectations for a more varied and flexible working environment and hours. These are macro level, but societal risks can have significant impacts at the micro level too—unique circumstances that impact one office, community, etc.
Technological	We have all seen the speed of technological advancement in recent years and these changes can drive significant risk into the organization. This category covers a huge area, from the rapid rate of change in software versions that shortens the period from launch to obsolescence, to the evolution of new platforms and capabilities on those platforms. This is one environmental factor that is often underappreciated as a risk category simply because the majority of risks are seen as positive (opportunities).

Similarly, elements of these risk categories are considered too insignificant to worry about—for example, a location in an area of seismic activity. This is a geographic risk that exists, but it is often completely ignored from a risk management perspective simply because the likelihood of anything more than a *minor inconvenience* occurring is considered extremely remote. That's fair enough, but even if there is only a 1 in 100 chance of a devastating earthquake in any given year, it's still a possibility, and the impact will be severe. If the company has ten such 1 in 100 risks, the law of averages says that one of them will occur every 10 years. Now we are starting to play dangerous games if we ignore them.

Of all of the environmental risk factors identified above, the only one that consistently gets active risk management attention is the area of competitive risks. Even here the management is frequently reactive rather than proactive. Organizations don't drive internal initiatives based on the possibility of a competitor taking certain actions; rather, they wait for a competitor to announce that they have the feature (or at least for rumors of it to emerge), and then they respond. Technically this is now an internal risk, and we'll look at those next. This approach can be a devastating strategy for the organization, and we don't have to look far for two recent examples.

In the 1980s and 1990s, Sony dominated the portable music market with the Walkman and then the CD Walkman. The name became synonymous with the product, and competitors struggled to gain a tiny share of the market. However, Sony didn't consider the risks of competition; they didn't see Apple coming, and when the iPod launched in 2001, Sony was virtually wiped off of the portable music player map. For Kodak, the situation was even worse. The company went from dominating film photography to bankruptcy because it failed to recognize how digital photography would change its market—despite being part of the invention of digital imaging.

We'll look at risk management approaches in much more detail later in the book, but I have no issues with organizations adopting a strategy of risk acceptance for most external risks—the conscious decision not to invest in active risk management because the return on the investment is not there. Consider the traffic signal example again—you can't influence when it changes, so why would you try?

However, that doesn't mean that the risks should be ignored because the impact will still be real, and you need to understand the consequences if the risk triggers—develop contingency plans, potentially alter business decisions to avoid exposing the organization to some of the risks, etc. This is where many organizations fall down, particularly on the less obvious risks. It's fairly easy to stay abreast of economic risks because the economy is an integral part of the information that we are exposed to every day as human beings, but what if a competitor is expanding in one of the cities that you have a manufacturing plant in? How confident are you that you will know that in time to plan for the potential loss of resources? If you do find out, will it be because of a conscious strategy to stay aware of your environment or through someone overhearing something or through reading an article by chance?

Generally speaking, organizations have considerable room for improvement when it comes to understanding and reacting to their external risk environment.

Internal Risks

In addition to the risk environment within which the organization operates, there are the more direct categories of risk that are driven internally. These categories of risk are affected by the organization's own actions and as a result are the ones that tend to get the most focus. These risks will likely be more familiar to you, and as is so often the case, they are almost exclusively considered in a negative sense. However, all of these can have opportunities (positive risks) as well as threats (negative risks).

Traditionally four categories of these business risks are identified: compliance, financial, operational, and strategic. Table 1.2 provides an overview of those categories along with an additional category that I have added—technological. The risks that an organization faces from

Table 1.2   Internal risk categories and descriptions

Category	Details
Compliance	Compliance risk is the risk to the organization from the need to comply with laws, regulatory frameworks, etc. Examples of the negative implications of a failure to comply are obvious: censure, exclusion from a professional body, perhaps even legal action. The positive elements are less obvious, but they are still there—being one of a few organizations able to claim that they have been given the highest level of industry recognition by the governing body for example.
Financial	Financial risks are the risks associated with investment decisions that the organization makes. Every proposed project involves a degree of financial risk—whether it is approved or rejected. Financial risks are often assumed as a result of some of the other risk categories, but managing these risks should be key to the decisions made around projects—does the expected return justify the investment that is being made?
Operational	Operational risks are those that stem from the day-to-day execution of what the organization does. This is a very broad category and may show itself through quality, customer service, productivity, employee satisfaction, or any number of other factors. For most organizations operational risks need to be broken down to a lower level to be properly understood and managed, the operations category is simply too broad.
Strategic	Strategic risks result from the directional decisions that the organization makes—the goals and objectives that it sets and the strategies and plans that it puts in place to achieve those goals and objectives. This is the most fundamental type of risk for the organization and will drive all of the others.
Technological	We looked at technological risks as an external environmental factor above, but there is also significant internal risk from technology. Decisions about which technologies to use can drive significant risks into the organization. If we choose to embrace new technology then we may face steeper learning curves, more teething problems, etc. If we instead decide to use older platforms then we may be faced with an earlier forced upgrade, lower performance, and reduced feature sets.

within—the risks associated with operating the business—will fall into one or more of these categories. While each individual risk may not be categorized into one of these buckets, it's important to understand the areas that drive risk within the organization. This will provide the organization with an appreciation for where it is exposed to threats and/or has opportunities that it may be able to exploit. However, we can't simply consider each of these as isolated factors; they combine to define the organization's overall risk profile.

The risk profile is simply a summary of the risks faced by the organization. It is not a risk management tool. It doesn't have enough detail for that, but it is a simple way to view the organization's risk exposure that can be used as an input to the corporate decision-making processes to ensure that decisions are taken with a complete, accurate, and current set of information. If we think of the risk exposure to all of the factors discussed as data elements in the process then the organizational risk profile is the tool that processes that data into actionable management information.

Later on in this section we'll look in more depth at the theories behind a risk profile, and we'll explore some practical tools for creating and maintaining the profile.

Risk Inevitability

Before we leave this overview and start delving deeper into specific risk elements, let's look briefly at the reality of risks. If we go back to our driving analogy, the only way to avoid the risk of having to deal with a changing traffic signal is to never drive anywhere with traffic signals. Most of us would agree that as a strategy that approach has a significant downside. In the vast majority of scenarios, we have to accept that the risk exists and that we may need to deal with it. If we eliminate the risk entirely (don't drive near traffic signals) then we may not be able to complete our functions as people—getting to work, running errands, socializing, etc., or we will subject ourselves to other risks—driving on more rural roads that are less well lit, have inferior road surfaces, fewer signs, or a greater chance for wildlife in the road. For most of us it simply is not practical to eliminate the risks presented by traffic signals.

The same is true for organizations; risk is not only inevitable, it is necessary. Those of you who have studied risk in the context of project management will probably have learned that risk elimination is a legitimate risk management strategy, and it is; however, it can only be used in some situations. You simply cannot eliminate all project risks without also eliminating the project itself.

At the organizational level, it is no different. Accepting a decision means accepting the risks that are associated with it. Elimination of one group of risks will result in additional or increased risk exposure elsewhere, likely with minimal impact on the overall risk picture. If the risks can't be accepted then the decision can't be made, but that is still only a transfer of risk elsewhere. For example, if an organization has $100 million to invest into the project portfolio in the next 12 months, then the expectation is that the $100 million will be invested. If a $20 million project is rejected because the risk/return calculation is unacceptable, then that $20 million needs to be allocated to other projects and the risks that are associated with them, or not invested at all with the risks associated with not being able to get the same level of potential return.

A commercial organization exists to make money and to do that it needs to make investment decisions that strive to maximize opportunities while minimizing threats—and that requires strong organizational risk management. Public sector organizations may not have the same profit driven goals, but they are still expected to deliver their services as efficiently as possible—doing the most for the lowest cost. That requires maximizing opportunities and minimizing threats—risk management.

In this first section of the book, we are going to focus on the foundations of risk management, culminating in the development of an organizational risk profile that will summarize the organization's risk capacity and risk tolerance. However, before we get there, we are going to need to understand a few risk-related concepts.

2

Risk Relationships

In the previous chapter, we looked at the different categories of risk from both inside and outside the organization. This gives us foundation knowledge, a basic understanding of the risk source, and potential impact on the organization. However, this understanding is still far too basic to be able to effectively manage the risks with any expectation of success. Effective risk management requires a detailed understanding of how the risks relate to one another; how they will respond to different management approaches; and how much time, effort, and money will need to be invested before a meaningful impact on the risk is achieved.

The first step is to understand how each individual risk and risk category interacts with others—the relationships between risks. As an example, think about a change that occurs within an organization—say the retirement of an executive. That single act will have a lot of impact—maybe a new executive will be brought in from outside who will want to bring some people with him or her and that will cause moves and changes. They may decide to reorganize, which will drive some other changes. Some of their staff may not like the changes and leave, creating openings for others to be promoted and in turn for someone to be hired to fill their old position. That one single act—the retirement of a senior individual—can create a cascading impact that ultimately results in the hiring of someone new in the mail room.

The same situation occurs with risks. A change in one risk can have a wide-ranging effect elsewhere in the organization, and if we don't understand that those relationships exist and the potential impact they may cause, then we will never be able to develop an effective risk management

11

strategy. There are two types of relationship between risks that we need to consider:

1. *Risk driven relationships.* In these cases the risk itself is driving associated risks. As one risk changes its profile, it drives change in associated risks.
2. *Action driven relationships.* In these cases, the actions that we take to try and control the risk drive changes to related risks. This effectively requires a compromise in our risk control activities.

Of course, both situations may exist for the same risk. In fact the risks that have the most risk driven relationships are often the most serious. Therefore, they are the ones that are in the most need of actions being taken, even if those actions themselves drive additional risk exposure. Consider also that the relationships are not always negative. By taking actions to manage one risk we may be creating or increasing an opportunity (positive risk) elsewhere, or we may be mitigating a related threat (negative risk).

Risk Driven Relationships

Let's start with an example of this type of relationship to help us recognize it. Suppose that an organization is having problems with a systems upgrade that will deliver new regulatory reporting—the system is failing quality assurance, and the schedule is being delayed. As a result there is a high likelihood that the organization will fail to make the deadline for the new reporting requirements (increased compliance risk). The regulator will then have the option to impose fines on the company for noncompliance (increased financial risk), lower the company's rating (reputational risk), and subject the company to increased monitoring and audit requirements (increased regulatory risk).

We can clearly see here that the impact of one risk becoming real can have a significant effect elsewhere in the organization and can cause problems that go far beyond simple problems with a reporting system. We need to understand these relationships and how the impacts can spread out from the central issue if we are going to make intelligent risk management decisions.

If we don't understand the connections between the risks, then we may not be that concerned about a slight delay in the schedule beyond the deadline. We may have a sense that not being ready by the time that the regulator requires the change to have been implemented is a bad thing, but understanding the full extent of the impact will help us to decide whether the delay is acceptable or whether we need to invest more in trying to avoid the risk from triggering.

It's also important to understand that the increased risk impact does not require the first event in the chain to become real. This can be a difficult concept, but it's key to understanding risk exposure. In the example that we used, let's look at nothing more than the levying of a fine if the deadline is missed. This is a fairly black and white situation—meet the date and don't incur a fine, miss the date and be penalized.

However, from a risk standpoint it's not that simple. If the fine is $100,000 and we originally estimated the likelihood of missing the date at 10%, then we needed reserves of $10,000 to cover this risk. If we do the project ten times, then we will miss the deadline once (10%) and incur a total fine for the ten projects of $100,000—$10,000 per project. The reason that we allocate reserves of $10,000 for the risk even though the actual impact will always be $0 or $100,000 is that the $10,000 contributes to an overall reserve pool for all of the risks—the reserve for the risks that don't become real problems offsetting those that do.

Suppose that challenges on the project require us to re-assess the likelihood of missing the date at 50%. Now we need reserves of $50,000. If we do the project ten times, then we will miss the deadlines five times (50%) and incur a total fine of $500,000 or $50,000 per project. We still don't know whether the project will deliver on time or not, but the risk profile has already changed. The reserve requirement has increased because of a change in circumstances on the project.

We'll look at these concepts of reserves in much more detail later in the book.

Action Driven Relationships

Anyone familiar with project level risk management understands the trade-offs that occur when we manage risks. We choose to invest time, effort, and money into various risk management strategies in order to reduce the likelihood of a risk occurring, the impact that the risk has on the project if it is triggered, or both—a trade-off of some managed impact now to reduce the chances of a more significant, less-controlled impact later.

Action driven risk relationships are an extension of this concept. They involve the acceptance of additional risk in one area to reduce the risks elsewhere. As a simple example, suppose that the organization in our regulatory reporting scenario is not prepared to accept the increased risk of late delivery of the new regulatory reporting. Instead they decide to move resources from other projects in order to try to recover from the delays and deliver on time.

Our project management training would tell us that this is a cost increase—we are sacrificing budget in order to try and preserve schedule, and for the regulatory reporting system project, it is indeed a cost increase. However, from an organizational standpoint the change is cost neutral. We still have the same number of people working on the portfolio; we have simply diverted some of them from one initiative to another.

From an organizational standpoint, this is a risk play. We are creating an action driven risk relationship by increasing the likelihood of problems with scope, quality, schedule, etc., in some projects as a direct result of trying to protect the schedule in the project that poses the most organizational risk impact. Generally speaking, we should be looking to generate an overall improvement in risk exposure—the total amount of risk that the organization faces as a result of creating these action driven risk relationships is less than the total amount of risk faced before we created the relationships, but that's not a hard and fast rule. In our regulatory reporting example we are likely to accept additional risk exposure if it means avoiding being in breach of the regulatory framework that we operate under—a case of the type of risk being important. There are other situations where we might consider a higher overall risk exposure if it means that we can move risks away from critical areas—protecting a revenue generating project at the potential expense of a project that is designed to drive operational savings, protecting costs by accepting increased schedule risk in exchange for reduced financial risk, etc. Your organization likely makes many of these decisions already, just based on a less structured, more gut-feel approach that we should move people from one project to another to try and keep things on schedule.

Managing Relationships

Risk driven relationships are predictable. We can follow the logic that says if event A occurs, then events B, C, and D are more likely to occur. This still requires detailed analysis. The connections aren't always obvious, but they do at least follow a logical flow. Managing these relationships consists of a proactive approach to understanding the relationship itself, the significance of the impact that the relationship can create (i.e., how substantial a problem can a triggered risk cause), and the work that can be done to manage those interconnectivities—either minimizing or eliminating the connection or ensuring that the impact is acceptable. This becomes part of risk analysis that we will look at in Section 2 of the book.

Action driven relationships are much harder to predict because they don't flow from the risks themselves, but rather from how the risks are addressed. As a result, a shift in the management approach can cause significant change in the action driven relationships—breaking some connections completely, creating new ones, and changing the impact in others. On the positive side, they are a lot easier to manage than risk driven relationships because we make conscious decisions to accept increases in risks 1, 2, and 3 in order to reduce risk 4. We control the impact.

Action driven relationships leverage work that is already being undertaken further down within the organization where project managers are undertaking risk management on their individual projects. This provides the organization with a solid understanding of the impact on a project of losing a resource for a period of time, diverting budget to another project, increasing scope, shortening the timeline, etc. Because most of our action driven relationships are a combination of these project level risks we can, as long as the risk management is effective, make quick assessments of the impact. This analysis is largely reactive from the organizational perspective. It is in response to a situation that requires us to take the initial action that triggers the action driven relationship.

Understanding the relationships is a vital part of an organizational risk management process. Our overriding focus, and something that will become a recurring theme, is to ensure that the portfolio delivers the organizational goals and objectives that have been established. If that means sacrificing one or two portfolio elements (projects) to preserve the ability to deliver the greater goals, then that's an acceptable strategy and that may be the impact of some of our risk management. In most situations, we will have limited capacity to add additional resources (people, time, or money) to the overall portfolio, and even if there is some capacity for these additions, there will generally be some delay between the addition and the impact being felt. As a result, organizational risk management frequently becomes about making trade-offs—choosing the least unacceptable approach, sacrificing the project with the least impact, or accepting the risks with the lowest chance to derail the portfolio. The only way that those decisions can be made with any degree of confidence is with a solid understanding of the relationships between the risks.

3

Risk Impact

So far we have looked at the different categories of risk that an organization faces, and we have briefly considered the relationships that exist or are created between risks. In this chapter we're going to focus on what happens when a risk becomes real—the impact of the risk. When we start to consider organizational risks, the potential span of the impact is far more significant than a traditional project level risk. We need to be able to understand the reach of that impact—not only will it influence the steps that we take if and when a risk does trigger, it's also integral to determining the appropriate management steps. If we don't understand the full scope of the potential impact, then we can't make an accurate determination of these items:

- The potential risk exposure that the organization faces, which is a vital part of determining how aggressive and extensive the risk management approach should be.
- The appropriate management steps that should be taken to try to control the risk and/or its impact.
- The required reserves that need to be put aside. We'll consider reserves a bit later.

In the introduction we looked at a definition of a risk, and we said that to be considered a risk, it had to "impact the project's objectives" if it occurred. We can replace *project* with *program, portfolio,* or *organization,* and the statement is still relevant. At various points in the book, we'll consider all of those entities, as well as the PMO in its role of support organization to the execution of the portfolio.

Now if impact is our determining factor on whether something is a risk or not, then we need to determine what constitutes an impact. For our purposes let's say that:

Impact is a noticeable effect that occurs as a direct and near immediate result of the trigger event.

In other words, if the event that causes a risk to become real results in a change in objectives, then that's impact. We can't insist that the change be measurable because the severity may not be known until later. Remember that the change can be positive or negative because risks can be opportunities or threats. Impact is generally considered to be money based (financial) or time based (schedule) but may also be more profound—an inability to achieve an objective at all or a change in the achievable quality.

So now we know what an impact is, and I hope that we know what our objectives are because otherwise we don't have much hope of knowing if they are impacted. However, before we start looking at the impact of risks at the various levels of the organization, we have to consider the concept of reserves.

Reserves are just what they sound like—things that we hold back in case we need them. For risk management we consider schedule reserves—time, and budget or cost reserves—money. Historically organizations have been reluctant to formally acknowledge reserves as a separate entry in the project plan, and so PMs have become experts at the black art of padding estimates without getting caught. More recently there has been better recognition that reserves are real and necessary and are more commonly included as part of the project planning process.

There are two distinct types of reserve that we consider, and both have time and money constituents. We'll look at them in more detail in Chapter 5, but for now let's look at some basic definitions:

- *Contingency reserve.* This is the reserve that is documented as part of the project plan and can be tied back to the individual risks in the project. Contingency reserve is calculated based on the impact and likelihood of occurring of each of the risks that exist in our initiative (we'll look at how to calculate it when we look at it in more detail in Chapter 5).
- *Management reserve.* Management reserve is designed to deal with those risks that were not predicted during the planning process—the *unknown unknowns* if you will vs. the *known unknowns* of our identified risks. Management reserve is usually a percentage over and above the contingency reserve and access to it requires senior level approval.

So with that understanding, let's start looking at risk impact at each of the levels of the organization.

Project Level Risk Impact

We aren't going to spend much time here. Project risk management is a well-documented aspect of project execution, and it's not our area of focus. However, because it is familiar to us, it is worth using as an example for the fundamentals of risk impact. Risks have an impact in one of two ways:

- They are triggered, in which case they become real and require action. This is the obvious area of risk impact, and it requires us to initiate the contingency plans that we have developed and utilize the reserves that we put aside for risks.
- They require management that redirects resources from project tasks more directly associated with the completion of project deliverables. This is less commonly considered as a project impact, but it is always impactful. The risk may or may not become real, but if we choose to actively manage a risk, then this cost will always be incurred.

At a project level, we allow for the first scenario by creating contingency reserves of time and money that are drawn down as risks become real. We re-assess the potential impact during the project, adjusting the reserves as necessary, and if we have done our job properly, then our completion date is approximately equal to the work schedule plus the schedule reserve, and our total cost is close to the projected budget for the work plus the cost reserve. There may be an allowance for a management reserve for unexpected or unpredictable risks, but these can be considered part of the overall reserves for the purpose of this discussion.

Now in the real world I recognize that not all projects (or perhaps more accurately, not all sponsors) allow for these reserves, or project managers bury them in padded estimates. However, whether we choose to acknowledge their existence or not, the reality of risks and their impact exists—risks that trigger will delay the project and generate cost overruns whether we have allowed for them in our plans or not.

The second scenario, the impact of risk management activities, is included as part of the regular project plan and work breakdown structure. It is work that we have chosen to perform as part of our overall risk management approach, having consciously decided that the predictable time and financial cost is worth the benefit that it is projected to deliver.

That benefit will either be a reduction in the impact in the event that the risk becomes real, or in the prevention of impact entirely by managing the risk to the point where it doesn't trigger at all.

In traditional project risk management, we tend to think only about the impact that risks have on the project itself. With an organizational risk management approach, we need to recognize that all of our projects and programs are connected together and that the impact of one risk triggering in one project can have a far-reaching impact elsewhere in the portfolio. We'll look at that in more detail as we look at program and portfolio level risk impacts.

Program Level Risk Impact

In some ways, the program is the most complex unit within a portfolio when it comes to risk impact. By definition a program is a grouping of a number of related projects, and a risk that impacts any one of the projects that make up the program has the potential to impact the program as a whole. At the same time, the program is downstream of the portfolio and may be exposed to the impact of risks from above, and of course there are also the risks that are generated from within the program itself. If this isn't enough, programs also introduce the concept of time, which adds another level of complexity. Not all of the projects within a program will follow the same schedule, and it is entirely possible that some constituent projects within a program will be impacted by a triggered risk from elsewhere in that program before they have even been initiated.

To control this many variables it's important that program level risk management focuses on the right areas of impact. The projects that make up the program will have hundreds, potentially thousands, of risks collectively, and a percentage of those risks will trigger. However, only a subset of those risks need attention from the program level; the focus needs to remain on answering the following questions:

- Does this risk have the potential to impact the ability of the program to achieve its overall goals? If so, is the risk management that has been put in place sufficient to manage the program and project level impacts?
- Is this triggered risk impacting the program's ability to achieve its overall goals. If so is the contingency that we are implementing sufficient to control that impact?
- Does corrective action have to be taken to restore the ability of the program to achieve its objectives after a risk has triggered?

Now obviously these questions are asked at different points in the risk management process, but if the answer is no, then there is no need for the program manager to get involved. The risk can be contained at the project level and left to the project manager to control. There may be a need to maintain closer monitoring of the specific risk to ensure that the situation doesn't change, but that is again an action that should be downloaded to the project manager to control in most circumstances, although at times program management may provide inputs to that monitoring process if they have better visibility into the potential exposure.

If the answer to any of the questions is yes, then there is program level risk impact, or at least the potential for it, and the program management function needs to take an active interest in what is happening. That impact can take a number of different forms:

- Impact on other current projects within the program. For example, a risk triggering on project A causes impact on projects B and C within the same program. This is the most common form of program level impact, and it makes sense. By definition a program is made up of a number of projects that have a common purpose, so it makes sense that there would be a number of relationships between those projects. Note that even if the impact is as simple as a risk triggering in one project affecting another project, we still consider this to be a potential program level impact.

- Impact on a not-yet-started project within the program. This is a special case that forms a subset of the group above and occurs when one or more of the projects impacted by a risk situation has not yet started. In this scenario, we aren't able to take any immediate actions because the project isn't underway, so instead we need to ensure that the impact is incorporated when the project does begin. For example, suppose that a vendor that is being relied on to deliver multiple elements of a program is late on the first major deliverable. This has potential impact in other initiatives and may result in more aggressive risk management, changes to contingency plans, and an increase in the contingency reserve to reflect the increased exposure. If one of the projects hasn't started, then these adjustments can't be made, but the potential impact needs to be captured at the program level for inclusion when the project kicks off (if the situation hasn't been resolved by then).

- Impact on the program itself. The impact of a triggered risk may not always be directly on the constituent projects within a program, but rather on the program itself. Suppose that a key supplier of support infrastructure goes out of business. Clearly, this is a major impact on the support infrastructure project within the program, but it may not impact any other project. They may be able to continue with project execution unaffected. It is only when those other projects are complete and need to be deployed that the lack of a support infrastructure will be a problem. However, the impact on the program as a whole is immediate—the program cannot be successful without the infrastructure and actions need to be taken to address that. Because the program is executed through projects, the contingency plan that the program implements is likely to drive changes into the projects that make up the program, and perhaps even create additional projects to address the issue if the impact is significant enough.

In addition to these project driven risk events, there will also be risks that are identified, analyzed, and managed at the program level itself. These risks don't naturally fit to any one project, but they still need to be managed effectively if the program is going to be successful. Frequently these risks are either missed or are not managed effectively due to a lack of resource assignment at the program level and/or the lack of formal risk management processes beyond the project.

Before we look at examples of program level risks, let's start separating programs and the portfolio from projects in terms of the role that they play within the organization. Projects are focused on deliverables—they are set up to deliver a defined set of requirements by a fixed date for a fixed budget and at an agreed-upon quality standard. In recent years, the concept of *fit for purpose* has started to creep in—the ability of the project deliverables to meet the business needs that they are intended to support, but that's still a secondary concern.

At the program level, our focus is broader than that. If the program is designed to implement a work automation system that will reduce operating costs by 10% after 12 months, then the program manager needs to be focused on that 10% target. Clearly, they still need to deliver the system that will be the vehicle for achieving that saving, but they need to be prepared to make decisions that compromise elements of the individual projects in order to protect the business benefits. At the portfolio level, the focus becomes even more strategic, and we'll look at that more as the book progresses.

With this focus in mind it is now fairly easy to identify some program level risks that don't fit into any single project—the risk that the system won't deliver the expected cost saving and the risk that the cost saving won't be achieved until more than 12 months after implementation are two obvious ones. In this situation the solution that we implement will be similar to the final bullet point above—we proactively drive changes into the projects that make up the program in order to address the business risk, even if that affects the functionality that the program is set to deliver.

These changes have the potential for considerable impact on cost, quality, resourcing, scope, schedule, and of course risk—all of the project constraints. In Section 2 of this book we'll look at the concept of the constraints hierarchy and how that can help us to determine which of these impacts is acceptable.

Another consideration that we have to make when we move from the project level to the program level is the increased uncertainty. There are now far more variables—the multiplying of risk because of the increased number of projects, the addition of the risks that are associated with the program itself and what it is expected to deliver, and simply the complexity of the relationship between all of these variables. This should lead to an increase in the management reserve as a percentage of the contingency reserve. By definition more uncertainty means a greater likelihood of unexpected problems occurring, and those problems still need to be accommodated.

Before we leave this section and move on to portfolio level risk impact, it's worth a reminder that while a change or triggering of a risk within a project that makes up a program has the potential for impact, it's not automatic. Look back at the definition of impact at the start of this chapter and use that as the guide for determining whether any program level impact has occurred, and hence whether any actions are needed. In many cases the impact is confined to the project where the risk has triggered, or where the risk exposure or management approach has changed.

Portfolio Level Risk Impact

It's tempting to view the portfolio as the consolidated view of all of the programs, but that's not true. As we saw in the section above, a program can include a number of projects that are complete, and others that have not yet started, whereas a portfolio is the grouping of projects that are currently underway or planned for the current cycle. We frequently consider a portfolio in conjunction with an annual planning exercise because that's often when projects are reviewed and approved for execution within the next fiscal year—effectively being added to the portfolio as they are approved.

Because a program consists of a number of related initiatives, there are elements of commonality between projects within the program, which makes it easier to understand the impacts of risk events. In a portfolio, the only element that the projects share is the fact that they are all being undertaken within the same budgetary cycle or time frame. There may be commonality among a subset of projects within the portfolio, but not across the entire portfolio. This requires us to take a different view of risk impact.

Before we dive into the details, let's take a step back and look at the basics of establishing a project portfolio. On a regular basis, generally annually, an organization will identify a number of objectives that it wants to achieve. Some of these may be mandatory—changes to a regulatory framework, for example. Some may be driven by business necessity—replacing obsolete equipment, upgrading buildings, etc. Some will build on previous years—phase two projects within a program. Finally, some will be aimed at achieving strategic priorities—reducing costs, increasing revenue, gaining market share, expanding into new markets, etc. The way that an organization sets out to achieve these goals and objectives is through projects—each individual project is tasked with achieving a set of deliverables, which will in turn contribute to the business goals and objectives. For example, a project to develop a new product will deliver a set of features that are designed to fill a gap in the market. By selling that product the organization will achieve its business objectives of increased revenue and perhaps increased market share.

Because the strategic priorities that drive the goals and objectives are generally reviewed or set on an annual basis, the approval of most projects often follows the same annual cycle. Therefore, it follows that the portfolio is often seen as an annual entity because from a structure standpoint the portfolio represents the entire set of projects that the organization is undertaking, However, a portfolio can also be an ongoing and evolving entity that drops projects as they are completed and adds projects as they are approved. (They become part of the portfolio before they are initiated because they are an approved part of the mechanism for achieving the organization's goals and objectives.) It doesn't have to be an annual snapshot. Similarly, some organizations will create separate portfolios within different areas of their business, especially if they have multiple stand-alone PMOs on a business unit by business unit basis.

It's important to understand the way that a portfolio within your organization is structured so that you have an accurate view of what it encompasses, but an organization should not manage its portfolio(s) that way. We talked previously about programs being more strategic than projects, and the portfolio takes it to a higher level again. The focus should be on the goals and objectives, not the individual project, or even program, deliverables. This is paramount when we consider portfolio level risk impact; we must always consider the impact on the business goals that the organization is seeking to achieve.

From a mechanical standpoint, you can view the process for determining whether there is portfolio level impact that is driven from projects or programs as identical to the process for determining whether there is project driven program impact that we looked at in the section above. There is no point in repeating those points here. Where we differ from program level impact is in the way that we determine whether there is any impact to the portfolio.

At the start of this chapter we defined impact as "a noticeable effect that occurs as a direct and near immediate result of the trigger event." That's still valid at the portfolio level, but the noticeable effect that we are looking for is on the business level goals and objectives that the portfolio is responsible for achieving. In some ways, this makes portfolio level risk impact easier to deal with than program level impact because the vast majority of risks have impacts that will never reach the strategic level. They may be impactful in the context of a project, but at the portfolio level, they don't even register. For example, if a $1 million project has a risk that triggers a $40,000 impact, that's noticeable—4% of the overall budget. However, if the project is part of a $250 million portfolio, then the portfolio manager will likely not even be made aware of the fact that the risk has triggered, because the overall impact on business objectives can be contained and resolved.

That's not always the case, just as every project has one or two single points of failure, so every portfolio has one or two *weak spots* where an impact cannot easily be contained and where even a relatively minor project level impact can affect the portfolio's ability to achieve one or more of its objectives. Typically this occurs in situations where there are only one or two initiatives within a portfolio focused on a particular objective, although it can also occur if the portfolio objectives are too granular— revenue growth by product, by quarter for example. That is ultimately an issue with the strategic planning and goal setting process rather than portfolio management, and we'll look at that later in this chapter.

It's vitally important to apply this *impact filter* at the portfolio level. The sheer number of projects that contribute to a portfolio will result in a large number of triggered risks, changing risk exposures, etc., and unless there is some way to focus on the risks that have the potential to affect the portfolio's ability to deliver business benefits, there will be a lot of time and effort wasted. It also needs to be noted that while there is often a close alignment between program management and PMO functions, portfolio management is another element that tends not to have a large number of resources. It is therefore necessary to ensure that those resources are applied in the most effective manner.

Where there is legitimate potential for portfolio impact, the portfolio manager needs to have a lot of freedom to address the situation. This chapter is on impact, and we will spend a lot of time looking at management later in the book. However, it must be understood that a lot of the actions that the portfolio manager takes are going to drive risk impact into the programs and projects that make up the portfolio. The portfolio manager will need to act by driving actions into those initiatives, and that inevitably means favoring the projects that are going to do the most to ensure that the at-risk business objectives are met at the potential expense of other initiatives. Even if this is simply a reallocation of resources, the programs and/or projects that lose resources will be facing increased risk exposure that needs to be assessed and acknowledged.

The portfolio will also be faced with its own inherent risks—exposures that are not driven out of the projects and programs but come from the strategic level within the organization. We have already identified one of them in this section, a reliance on only one or two projects to deliver a business objective that leaves minimal flexibility in the event of problems. As a strategic function within the organization, virtually any of the business level risks that we looked at in Chapter 1 could create risks for the portfolio, but some of the more common risk areas include:

- *Project mix.* If the portfolio is skewed heavily toward one particular area, then this can create significant risk by reducing the available options for the portfolio manager. Similar to the adage of *putting all of your eggs in one basket*, if the portfolio is focused heavily on one particular objective, then risks that threaten that area will jeopardize the entire portfolio and offer limited chances of recovery. This is a risk that can be largely avoided by ensuring that the organization has a broad mix of strategic priorities and hence goals and objectives, but this is not always possible, especially in smaller or newer organizations.

- *Inappropriate goals.* Organizations often demonstrate a lack of objectivity in determining the goals and objectives that they establish for a given period, and this can result in unrealistic expectations. If the portfolio manager is faced with trying to achieve goals that are simply not achievable, then not only will a lot of time and effort be put into striving after the impossible, there will be significant collateral damage as realistic objectives that were deemed less important are sacrificed. It's important to ensure that there is some validation of objectives before they become set (the portfolio management function can assist with this).

- Excessive goals. There is only so much that can be achieved within a portfolio and if the organization sets too many objectives, then the portfolio manager can be forced to choose which objectives to protect and which to sacrifice. Those decisions should have been taken during the planning and project approval phase. This issue can also be the result of objectives that are too detailed or granular. It's important to have a plan that breaks out to lower levels of the organization, but that shouldn't be managed at the portfolio level, because it removes the flexibility that is required to manage effectively.

Clearly these drive from the organization, which is what we would expect with the strategic function of portfolio management, but there is also the potential for risk impact at the organizational level, and we will consider that next.

Organizational Level Risk Impact

The organization is clearly the highest point in our pyramid, so everything that happens at the project, program, or portfolio level will be felt by the organization, and from a practical standpoint we will consider the portfolio as the highest point for risk management. However, it's important to understand how risks impact the organization because that risk will directly drive a requirement for risk management either into the portfolio, into daily operations, or both. If the operational impact is significant enough, that may also lead to changes in the portfolio as new projects are approved to implement the required operational changes.

For the organization, impact will come either from the various external and internal risk categories that we considered in Chapter 1, or they will come up through the project execution organization—the projects, programs, and portfolio.

If we consider the external and internal risks first, they will create impact in a number of different ways:

- *Organizational goals and objectives change.* Organizations are constantly evolving, reacting to circumstances, and proactively seeking out new opportunities (positive risks remember). These shifts in the organization and/or its environment may result in a conscious decision to alter the strategic priorities, and by extension the expectation, for the goals and objectives that the portfolio is expected to deliver. Examples include a need to shift focus from revenue generation to cost saving as a result of economic events. There may be an opportunity to expand into new markets rather than consolidate within existing markets as a result of a competitor's failure, etc.
- *Tangible deliverables change.* Sometimes shifts in the risk environment may cause the organization to adjust the way that it seeks to achieve its objectives. To use an example with a positive risk for a change, suppose that there is an unexpected opportunity to acquire a new building. This may result in the cancellation of a planned construction project to create more usable space within an organization's existing premises and the initiation of an office move. However, the underlying objective remains the same—to create space for more employees.
- *Funding changes.* This may be an impact from the previous two points. For example, if an opportunity to acquire a competitor arises, then that may result in the availability of new funds to expand planned work and create new projects associated with the acquisition. Alternatively, that same acquisition may tie up capital and reduce the portfolio budget. Funding is generally considered to be a reactive impact—it's driven by something else that causes the organization to change its investment levels. That is usually true at the organization level, but if your organization has portfolios assigned to different department level PMOs, then funding impact can occur simply through a rebalancing of the investment across the various portfolios.
- *Timelines change.* This is another example of a typically reactive impact—one driven from other impacts. For example, if we learn about a competitor who is planning a new product, then in order to beat them to market we may decide to accelerate the development and release of a product that we were originally scheduling for later in the year.

It's clear to see that all of these will drive significant change into the portfolio, and on into at least some of the projects. These are generally the most common types of organizational impacts that we consider—the ones that generate a top-down impact into the portfolio that likely cascades out into a number of programs and numerous projects. However, we also have to consider the potential for risk impact at the organizational level that is driven up through the portfolio.

To consider this type of impact we have to look back at the earlier sections of this chapter. When the risks from the project, program, and portfolio level cannot be absorbed within the portfolio, they create organizational impact. We said earlier that the portfolio exists to deliver the goals and objectives of the organization. To achieve that, portfolio management should be given a fair degree of flexibility to make changes within programs and projects—compromising some areas in order to protect the overall goals. However, at times the portfolio simply won't be able to achieve what it is expected to deliver. We looked at some of those risks in the portfolio section, and they can be managed through adjustments to expectations or adjustments to portfolio constituents. Management doesn't always work however, and some risks will trigger. The impact of some of those triggered risks won't be able to be contained within the portfolio. Additionally, the portfolio is an extremely complex entity with numerous variables, and there is a significant chance of an unidentified risk becoming real, resulting in an impact that is much harder to contain. Sometimes simply throwing more money, more resources, or more time at the problem doesn't work.

Whenever a risk triggers within the portfolio that makes it impossible to achieve one or more of the goals and objectives, we have an organizational level impact, and the organization needs to determine how to contain the impact.

PMO Level Risk Impact

Before we look at the PMO in too much depth, let's consider exactly what we mean by the term. When most people refer to a PMO they are generally talking about a project management office—a function that exists to support the projects that are being executed by the organization or one part of the organization. IT is a common department for a PMO to appear.

However, PMO can also refer to a program management office— sometimes seen as the same thing as a project management office, and sometimes specifically to manage a particularly large program within the organization.

More recently we have seen a trend toward EPMOs—enterprise project management offices. These are attempts to consolidate multiple PMOs (either usage) into a single organization-wide entity.

For the purposes of this book, it doesn't matter which particular variation your organization uses, or the specific functions of your unique PMO, the concepts will remain the same.

A PMO function is a service provider to the project execution process; it provides a support infrastructure that may include methodology and process, tools and templates, expert guidance, audit and control, centralized tracking and reporting, etc. As such, it is not directly involved with the day-to-day risks and impacts that we have considered, but it still remains a key stakeholder that may be indirectly impacted by those risks—through the provision of subject matter expertise and guidance to try and develop and implement contingency for example.

The area where the PMO has the most significant impact however is in understanding what caused the risk impact and what needs to be done to improve the way that projects are executed as a result. This is the most crucial impact of any risk because it is the one that can help to prevent future problems from occurring.

Every risk that is triggered provides the organization with an opportunity to learn and grow, and the PMO is responsible for ensuring that growth occurs:

- Analyze the risk event that occurred, the proactive actions taken to try to manage the risk, and the reactive actions taken to try and contain it (or expand it if it was a positive risk).
- Understand the underlying cause of the trigger event and consider whether the actions taken were appropriate.
- Review existing process, tools, training, etc., to identify areas where these can be improved to better manage similar situations in the future.

This in turn will lead to the implementation of changes and the monitoring of those changes to ensure that they delivered the benefits that were expected.

There may be occasions where inadequate process or insufficient training are the reason why risks were triggered, but these will be the exception in a well-managed project execution approach. The more normal scenario is one where project execution is constantly refined through a process of evolution to react to the organization's ever-changing risk environment.

Impact Containment

Recognizing that most of the risks that we deal with have the potential to cause problems rather than solve them (they're threats, not opportunities), once a risk has triggered, containment can be difficult. The event can't *unhappen*, and the focus shifts to trying to minimize the extent of the impact at multiple levels of the organization. The PMO can try to improve project execution for the next time something similar happens, but that doesn't help us now.

Containment is driven by contingency plans that are developed ahead of time and are executed once a trigger event occurs and the risk becomes real. The use of contingency is itself an exercise in risk management because it diverts time, money, and/or effort away from other areas in the project, program, or portfolio and directs them to minimizing the risk impact. This may well be appropriate, but we have to ensure that the cost of the cure isn't worse than the impact of the risk. We'll see in Section 2 that this analysis is conducted when the contingency plan is developed, but we also have to review the real impact to ensure that it is at the level of severity that we expected so that our containment efforts are appropriate.

4

Risk Command and Control

At a project level we talk about risk management as a general concept—a knowledge area to use a PMI term. We then consider specific processes like risk identification, risk analysis, risk prioritization, etc., that fall within that knowledge area. That works well for projects, but when we are considering risk from higher up in the organization, that's not enough. A portfolio has more variables, each of which can not only be a source of risk but which can also impact the risks to which the portfolio is exposed.

That's not to say that our traditional risk management processes aren't valid—they are, and in Section 2 we will develop an organizational approach that is based on those processes. This helps to provide consistency between the portfolio, programs, and projects, at least when it comes to common language and broadly similar approaches. We add to that the concept of an organizational risk profile, and we'll develop that in the next chapter as we transition from theory into practical application.

The command and control structure provides us with some guidelines that will help us to govern how the risk profile and the risk processes are applied—consider these as the operating principles to be used in organizational risk management:

- We require an accurate view of the organization's risk exposure.
- We require an understanding of the organization's current ability to absorb and withstand risk events.
- We need an understanding of the overall accuracy of the risk analysis that has been completed.

- We need to monitor the appropriateness of the risk management approaches.

Collectively these form a command and control structure that supports and manages the overall approach to organizational risk management. They provide the portfolio manager (or whichever portfolio level resource is tasked with organizational risk management) with the *filters* that they have to apply to the pure processes in order to have an accurate understanding of the true risk situation—the tools to manage the application of the risk management processes on their portfolio.

Let's review each of these items individually.

Understanding Risk Exposure

Risk management processes are designed to identify specific risks early on so that they can be analyzed, prioritized, and managed appropriately. That's not the type of risk exposure we are discussing. Instead, we are considering organizational risk exposure—at a high level what does the risk environment look like. In Chapter 1, we looked at the external and internal risks that the organization interacts with on a day-to-day basis. Some of those risks will impact the company, and similarly the actions of the company will impact some of those risks.

When we approve projects that make up a portfolio we are making decisions that impact those risk relationships, and this element of command and control requires us to understand those impacts. Consider a fairly simple example where we decide to invest a large percentage of the portfolio into the development of a new product offering. Before we consider the specifics of the projects that are going to contribute to that product, we can make some assumptions about the risk environment:

- We are going to reduce the risk of a competitor beating us in that product category because we are focusing heavily on that area and will probably gain some competitive advantage as a result.
- We may well be increasing the risk of a competitor gaining an advantage on us in other product areas. By narrowing our focus onto a major new product, there will be less attention paid to other product areas that may not have any impact on what a competitor chooses to do but will likely reduce our ability to respond to any actions that a competitor does take.
- We are reducing the diversification of the portfolio which may lead to increased financial risk—we aren't putting all of our

eggs into one basket, but we are reducing the number of baskets that we are using.

There may be others—increased operational risk as the products launch due to less familiarity, more market share growth opportunity (positive risk) if we are expanding into an underserved niche, increased regulatory risk if the market or product is regulated. The list can go on, but you can see that these changes in our risk exposure come as a result of the decisions that we make, and before we ever start executing on the projects.

It's important to understand these changes in risk exposure because they add weight to the items in our risk management processes—they help us to focus more closely on risks that impact our areas of increased exposure and be more tolerant of risks that impact areas where our exposure is reduced. These shouldn't be seen as either good or bad, they are just shifts in the natural risk exposures that the company faces. In times of aggressive expansion and growth, the changes may result in an increase in the overall risk exposure as we are prepared to take a few more risks in order to gain an advantage—a start-up or a small niche player trying to become mainstream are examples. By contrast, in times of consolidation or conservative planning, the overall risk exposure may be consciously decreased—reducing the potential upside but also reducing the downside. We often see this approach in times of economic uncertainty.

Ability to Withstand Risks

Risk exposure is only one half of the picture; we also have to consider how well the organization is positioned to withstand those risks. A 10% increase in the total risk exposure of a portfolio will be viewed differently if there is 20% overcapacity to absorb risks than if the portfolio was already struggling to deal with the risks that it was facing.

So how do we determine this capability to withstand risk? Let's look at an example that we can relate to from our personal lives. All of us have to balance our budgets at home every month. Our household has income from one or more sources and from that income we have to allow for:

- Regular monthly bills—mortgage or rent, utility bills, groceries, etc.
- Other regular expenses (things that might not be monthly but are still predictable)—house and car insurance, property taxes, clothing, discretionary spending, etc.
- Future planning—savings, retirement plans, college funds, etc.

- Emergency funds—money that we put to one side in case we need it, such as unexpected car repairs, vet's bills, etc.

Now let's compare those elements to a portfolio—we are looking at money in our household example, but the concept works equally well with time or effort.

Our regular bills, both monthly and otherwise, represent our project and program expenses—the cost of running the portfolio that is incurred on a regular basis—resource costs, office expenses, etc., and the less frequent but still predictable costs like payments under contracts with suppliers.

Our future planning is the money that we are investing in risk management—we are actively investing today to make the future more manageable.

Our emergency funds are our contingency reserves—the money that is specifically put aside to handle risks that become real on our projects.

In our personal lives we will hopefully have a little bit of money left over *just in case*, and in the portfolio we have the concept of the management reserve, the additional reserve that is put aside for dealing with the *unknown unknowns*.

This is where we look at the capacity to withstand risk. Determining how much management reserve is needed will be covered in Chapter 5, but it's a difficult balancing act—too little and risks can derail the entire portfolio, too much and we face lost opportunities to achieve greater returns from the portfolio.

Each time that an unexpected risk becomes real, we need to draw on our management reserve to deal with the problem. That also requires us to re-assess the portfolio in the context of the remaining management reserve—is the reserve still appropriate for the outstanding work on the portfolio (or program) and if not, what needs to change? This is no different from the approach we take in our personal lives. If we have a large cost that we didn't see coming, then we may have to cut back on discretionary spending for a while until we are ready for another *rainy day*.

Ability to withstand risk and risk exposure are the two cornerstones of risk command and control, and they will factor heavily in the organizational risk profile that we are going to develop in the next chapter. The rest of the items in this chapter are support functions to these two cornerstones, but they are still vitally important to understand.

Risk Analysis Accuracy and Currency

Our contingency reserve is based on the known risks that have come from our risk identification and analysis. It follows that the reserve is only as accurate as that analysis. If we fail to identify risks, or inaccurately assess the impact or likelihood of occurrence, then our calculated contingency reserve will be inaccurate. Similarly, if we fail to identify when individual risk conditions change, then we will not maintain an accurate view of the risk situation on our portfolio or its constituent elements.

Our organizational risk management processes will provide us with a framework to review and improve the risk analysis as the execution of the portfolio proceeds, but our command and control view of risk management requires us to understand how the current situation requires us to adjust our view of risks. For example, let's look at three separate risks that the portfolio is facing:

1. A risk that has been actively managed as part of a project for the last six months. It has responded well to the management approach and has slowly reduced in terms of likelihood to occur and impact if it does occur. It is still being managed by an experienced resource who monitors it closely.
2. A risk that was identified three months ago during the project approval process. It was analyzed in-depth to support the business case, but because the project has not yet been initiated, no one has looked at it since.
3. A risk that has just been identified and a couple of experts have offered the opinion that "it could be big."

Hopefully you can see how these three risks shouldn't be viewed as equivalents, even if they are all nominally subject to the same organizational risk management approach. Risk #1 sounds like it can be taken at face value, it's got a strong and consistent history and is being well managed—nothing to be concerned with there.

Risk #2 seems to have been analyzed fairly well when it was first identified, but it looks like nothing has been done with it since. At the least, we need to validate that nothing has changed before we rely on the information we have. We might decide that we can take some preliminary management actions, but we need to be prepared to adjust them if further analysis reveals that the situation has changed. There may also be an opportunity to review and improve the way that our risk management process has been applied—even though the project hasn't yet started. If there is risk exposure that can impact the portfolio, we may have an opportunity to at least monitor what's happening.

Risk #3 is our classic knee-jerk risk—it's got the potential to be disastrous, so we have to do something now even if we aren't sure what that something should be. In some circumstances that may be appropriate, but we need to be suspicious of anything that we hear about this risk until more work is done to quantify the situation. As we look to control and manage the risk, we should prioritize understanding and minimize unguided actions.

Looking at these in the detached environment of reading a book may make them seem obvious, but within the stresses of a portfolio, it can be easy to forget to apply this interpretation element when we make decisions based on the risk analysis that has been completed.

Appropriateness of Risk Management Approaches

The final area that needs a command and control approach is reviewing the management actions that are being taken in response to risks. There are four basic approaches to risk management of negative risks that can be implemented, and they are fundamentally no different at the project, program, or portfolio levels of the organization:

1. *Mitigation.* This is the most common approach and involves carrying out work now to try and reduce the likelihood of the risk occurring, the impact that the risk has on the initiative if it does occur, or both. Examples of risk mitigation are assigning a more experienced resource to a complex task in order to reduce the likelihood of it going wrong, or backing up data to an offsite location to minimize the impact of a data center failure.

2. *Elimination.* This approach is the complete removal of the risk and is usually taken when the impact is considered too severe. Risk elimination is rarely straightforward and usually results in changes having to be made to the way that the project is executed, often creating additional costs (time and money) and risks in the process.

3. *Transference.* Transferring the risk is the movement of risk exposure to a third party. It differs from elimination in that the risk still physically exists, but now someone else is responsible for it. Transfer may be full or partial, but generally only works on the financial element of a risk. For example, a fixed price contract will transfer the risk of a cost overrun to the vendor (because the customer pays a fixed price regardless), but the customer is still exposed to time delays if the product is late, quality issues if it does not meet the standard, or usability issues if it is hard to understand or poorly documented.

4. *Acceptance.* Accepting a risk is the decision to do nothing to try and actively manage the risk. This is not the same as ignoring

the risk; the decision to accept is consciously taken, and the risk should still be monitored in case the situation changes. Acceptance can be a valid strategy in a number of different situations—it can be used when the risk is not significant enough to require active management, or when the cost of managing the risk is not significantly different to the impact if the risk is triggered. It may also be used when the available actions simply won't be able to do much to reduce the likelihood or impact of the risk triggering.

With the management appropriateness element of risk command and control, we are concerned with ensuring that the right strategy is being employed. This is an area where our risk management processes should have significant control over the decisions that are taken on a day-to-day basis. They are developed and implemented to deal with the shifts that will naturally occur during the execution of the portfolio and its elements, and if a command and control view interferes with that, then at best we are causing confusion and at worst we are undermining the authority of those people executing the risk management processes.

Instead, this element of command and control is concerned with dealing with changes in the organization's risk exposure and/or in the ability to withstand risk—the two cornerstones of command and control that we looked at earlier in the chapter. When one of these environmental factors changes, we may need to rapidly change the way that risks within the portfolio are managed—shifting the management approaches in certain areas or changing the overall time, effort, and money set aside for risk management activities.

Think of the command and control function as an early warning system for the risk management processes—it is outwardly focused, looking for signs of environmental change, and then it drives wholesale actions when they are needed. At that time, the risk management processes and their owners become the vehicle for implementing the required changes.

Effective Command and Control

To be effective, risk command and control needs complete visibility. It needs to have insight into the risks that impact the portfolio and the organization at large, along with the changes that occur in that environment. It also needs to be able to monitor the risk management that is occurring within the portfolio so that there is always an accurate and current view of the overall risk management situation.

This requires the organization to remain proactive in the way that risks are managed at all times. Similar to how project managers look ahead to ensure that barriers to team progress are removed before they impact the project work, the person responsible for risk command and control within the portfolio constantly looks ahead for signs that risks are becoming more severe, that management is failing, or that early warning signs of a potential trigger are starting to appear. The ability to identify the early signs of trends and differentiate them from one off anomalous situations helps ensure the effectiveness of risk management. If this is successful, then risk management can always be proactive—adjusting to problems and taking advantages of opportunities before they become obvious. If this is unsuccessful, then portfolio resources will spend a lot of time chasing risks to try and minimize the damage, managing the wrong things, and generally wasting time, effort, and money.

Ultimately, this will be the single most significant aspect of risk command and control—a perfect mechanical implementation of the approaches that we will explore in this book will not work if they are executed without an understanding of how to maximize the benefits to the organization.

5

Creating an Organizational Risk Profile

In the first four chapters of this book, we looked at the concepts of organizational risk, how those risks are related to one another, how the impact of risk is felt at different levels, and some basic risk command and control concepts. Now we are going to begin to transition to the next stage. In the next section of the book we will develop our organizational risk management process and consider how it is applied at each of the different levels within the portfolio. However, before we get to that point, and as a transition out of this first section, we need to consider the concept of an organizational risk profile and how to develop one.

Let's start by defining exactly what I mean by an organizational risk profile. It is:

The summary of the risks that the organization is exposed to, combined with the risks that the organization's leadership has chosen to accept through the way the business is run, compared to the ability to influence, control, and absorb those risks.

The risk profile is important because it is the foundation that all portfolio level (and below) risk management decisions should be based on. Every organization has a risk profile, but most will choose to ignore it in making its decisions. The approach that I am outlining here, and the template that we develop, does not create a risk profile for the organization, rather it provides visibility to the profile that already exists.

Theory of the Profile

In order to understand the risk profile in more depth, let's break the definition into three different sections and then explore each section in more depth:

1. The summary of risks to which the organization is exposed.
2. The risks that the organization's leadership has consciously chosen to accept.
3. The ability to influence, control, and absorb those risks.

Risks to Which the Organization Is Exposed

This section represents all of the external factors that create or modify risks within an organization. We looked at these external risk categories in Chapter 1 and saw that many of these are beyond the control of the organization. However, that doesn't make them any less real. If the organization does not have a good understanding of the risks that are created by its operational environment, then it will become insular—operating within a silo and leaving itself exposed to events for which it is unprepared.

External risks are likely to have a higher percentage of situations where the management strategy is one of acceptance, either because there is minimal ability to influence the risk (a global recession for example) or because the cost of risk management is too high—the only way to reduce the threat of a hurricane impacting the company's property is to move the premises to a part of the country that is not susceptible to hurricanes (and that may well expose the organization to different environmental risks).

Because external risks can be hard to both quantify and to manage effectively, they are the ones that are most often ignored. This is an extremely reckless approach because the fact that they can be hard to manage demands an effective contingency plan. If you can't do much to prevent the risk from occurring then you need to be prepared to deal with the consequences and recover as quickly as possible. Additionally, external risks have the potential to be major drivers of the organization's overall behavior—it is a bold CEO who chooses to ignore the activities of his or her competitors. For a recent example of that consider the executives at Research in Motion, manufacturers of the BlackBerry. They failed to acknowledge the threat presented by first Apple with the iPhone, and then Google with its Android platform, and they may never recover.

External risks are generally relatively slow to evolve and therefore don't need to be monitored on as frequent a basis as internal risks, although organizations need to be careful in applying the same approach to all categories of external risk. Our example of competitors again demonstrates a situation where things can change quickly, especially in a market where there are many competitors and/or the industry is fast paced—such as mobile communications.

Risks Consciously Accepted

These are risks that are internal to the organization, and we considered the categories of these in Chapter 1 as well. These risks are accepted by the organization through the decisions it makes and the actions that it takes, and that acceptance should be a conscious one based on an understanding of the potential exposure. In reality, decisions are often made with only a rudimentary understanding of the risk impact, and frequently there is simply an understanding of relative risk exposure rather than absolute exposure. We know that if we try to get a product to market six months earlier, we will increase the risks; but we aren't really sure by how much.

A greater percentage of internal risks will be capable of being actively managed, although a decision to accept the risk is still perfectly valid. For example, we can mitigate the risk of causing operational problems from a new computer system by phasing the rollout of the system over a period of several weeks. However, we may still choose to go with a *big bang* style deployment and trade off the potential risk for the potential gain in operational costs that an earlier successful deployment will deliver.

Internal risks tend to be much more fluid than external risks, in large part because they are directly connected to the business. As the business situation changes so the risks themselves change. This means that they need more frequent monitoring and will likely be subjected to more frequent changes in management approach.

While organizations are generally not in a position to control external risk exposure, they have nearly total control over the level of internal exposure that they are prepared to accept. There may be a need to comply with regulatory frameworks—financial institutions have to comply with minimum capital asset requirements for example, and there is always a need to justify decisions to a board of directors and shareholders/investors. However, the organization can largely determine for itself whether it wishes to adopt risky strategies or conservative ones.

Ability to Influence, Control, and Absorb Risks

This third part of the risk profile is concerned with the organization's capability for handling the risks to which it is exposed. Understanding that every risk the organization faces can be managed, if only through acceptance, is only part of the picture. We also have to ensure that there are sufficient resources available to implement those risk management strategies. If we consider a personal situation, insurance is an example of risk transference—we transfer some of the financial risk of owning a house, driving a car, etc., by buying insurance. However, we can only have that insurance if we continue to pay the premiums. If we can't afford the cost of the insurance, then we can't transfer the risk.

The same is true with organizations. Every management strategy comes at a cost, and while we may decide that mitigation is an appropriate strategy to deal with a particular risk, if we don't commit resources to implement that strategy, then at best we have acceptance and at worst we ignore the risk completely.

Absorption is a slightly different concept and speaks to the ability of the organization (or portfolio, program, etc.) to absorb the impact of the risk if it triggers. Go back to our personal example of insurance. We may not think we can afford the premiums on our house insurance, but we are probably certain we cannot absorb the financial impact of losing our house to a fire or similar catastrophic event without insurance—suddenly the cost of insurance is manageable.

With organizations we look at absorption in terms of our reserves. Have we put sufficient resources aside to cover the contingency and management reserves for the risks that will become real? If the answer is no, then we need to be prepared to pull financial or people resources from other areas of the business to cover the exposure. If our risk planning is accurate, then they will be needed!

In the worst case scenario, there are insufficient resources within the entire organization to cover the exposure, and that should drive changes to the other elements of the risk profile. The company is risking its existence otherwise. If the potential risk exposure exceeds the totality of the resources that are available to the company, then it may find itself in an unrecoverable situation—much like our example of losing our home to fire without having insurance.

Building a Risk Profile

Before we can start applying these theories in practice we need to briefly discuss the level of granularity that the risk profile goes to. This is an organizational tool, and it needs to stay at a high level. A program will likely only be one entry in the risk profile, and most projects won't have a mention other than as part of the total portfolio. This is fairly easy to achieve when we consider external risks because they are organizational level anyway, but it can be tempting to get into additional details for internal risks. We need to avoid the temptation of increasing granularity or the profile becomes unmanageable, and we are unable to make appropriate decisions at the organizational level because we are too focused on details.

For the same reason we have to be careful how granular we get when determining the numbers that we are applying to the profile. This is particularly true when we consider the management costs of external risks; an example might help to demonstrate the point. Suppose that the organization has had some bad press and wants to manage its reputational risk exposure. The organization will spend money sponsoring events in its local community, making charitable donations, etc., and all of these will assist with boosting its reputation (lowering reputational risk). However, we can't say that the entire budget for these activities is risk management; the investment also contributes to marketing, for example. We therefore have to determine the percentage that we will use—perhaps it's the increase from last year's budget as the need to improve the reputation is driving the increase, maybe it's different. We can't get stuck on these details, so we need to make a reasonable assumption, document it, and move on to the more important aspects. Remember this is an organizational profile, and a few thousand dollars here or there shouldn't have a tangible impact (but that doesn't mean that we can replace reasonable assumptions with wild guesses).

Okay, enough of the theory, let's build a risk profile, or at least the structure of one. Table 5.1 shows my template for an organizational risk profile complete with a few simple examples. Don't worry too much about the details for now—we'll reference different parts of the profile as we work through this chapter.

Start by looking at the column headings, and you can see that we include both management cost, the cost that we know that we are going to incur through our attempts to control the risks that we are facing; and impact, the cost that will be incurred if a risk triggers. In both cases we consider the actual dollar cost and the resource (effort) cost. In this example, I have stated resource cost in terms of effort months or person months, but full time equivalent (FTE) can also be used where 1 FTE = 12 effort months/1 effort year, if that's more familiar to your organization. We also include a column for the likelihood of the risk occurring—the chance to trigger, and we express that as a percentage—10% chance to trigger means that it will occur once in every ten situations, 20% means that it will trigger twice in every ten situations, etc.

Table 5.1 Organizational risk profile template (partially completed)

Category and risk	Management Cost ($000s)	Management Cost (effort months)	Chance to trigger (%)	Impact—Financial ($000s)	Impact—Resourcing (effort months)	Owner
External Risks						
Competitive						
Competitor 1 launches before us	0	0	10	1,250	0	VP Products
Competitor 2 addresses performance issue	0	0	60	100	2	VP Products
Economic						
Recovery from recession slows	0	0	20	2,000	100	CFO
Political						
Proposed new environmental laws passed	20	2	70	250	12	Counsel
Internal Risks						
Financial						
Product expansion program	50	4	50	1,000	16	Program Manager 1
Operational efficiencies program	70	5	25	600	24	Program Manager 2
Rest of portfolio	80	6	75	600	12	Portfolio Manager
Operational						
Customer service issues	40	6	20	1,000	20	Operations Manager
Management Reserve				500	24	
EXPOSURE TOTALS	260	23		2,560	80.6	
Risk Capacity	300	24		2,500	84	
SURPLUS (SHORTFALL)	40	1		(60)	3.4	

You'll see that every risk has an owner. I've used generic job titles here for the sake of illustration, but you should make sure that every item has a name against it so there is no doubt who owns that risk. That person should have a more detailed record of the risks that are identified here and be able to provide more details if needed. They will also be responsible for monitoring the risk that they own and communicating any material changes.

In the case of project/program/portfolio related risks, the risk management plan for the various initiatives will provide that additional level of detail that the risk owner manages. However, in operational or external risks, there may be a less *risk specific* document—a competitive analysis or an economic outlook. In today's connected business, all of these documents should be accessible online and so links can be added to the profile to allow users to drill down as long as the user has the right level of permissions.

Linking to the underlying document also helps to ensure that the content remains current, although the summary numbers still need to be managed and updated whenever there is a material change. Regardless of who physically makes the change to the file, the responsibility for providing updated information sits with the risk owner, and they should have a defined schedule for those updates—weekly or monthly for internal risks, probably monthly or quarterly for most external risks unless something potentially significant is occurring. That schedule should be viewed as the maximum period between updates (even if the update is *no change*). Risks rarely impact us on a convenient schedule, and unplanned updates may well be the norm in maintaining the risk profile. If it isn't up to date, then it isn't helpful.

Understanding the Numbers—Risk Management

Before we study the profile to understand what it is telling us, we need to understand how the calculations are carried out. Let's start with the numbers in the two management cost columns as those are the easiest to understand. The entries in these columns for each risk category are simply the amount of money and the number of effort months that we have decided to invest into trying to control the risks. This should be an estimate of the amount of time and money that is being spent on risk management activities, something that is fairly easy for the program and portfolio sections as it will be the sum of the costs for the risk management activities in our project plans. It's a little harder elsewhere, but it should still be possible to calculate based on the amount of time and money that is being spent on different activities.

To get the exposure total for these entries (the 260 and 23 in the Exposure Totals row for the two management columns in Table 5.1) we simply sum the entries. The calculations here are straightforward because these are known costs—they are the conscious actions that we have decided to take in order to attempt to manage the risk exposure. If we have decided to use the acceptance strategy, as is the case with many of the external risks, then those costs are zero or close to it. The political risk that we identified has a small amount of effort and money—perhaps a small amount of internal lobbying and some well-placed political contributions.

The costs tend to be higher for internal risks—the diversity of the initiatives within the portfolio mean that the risks are many and widespread, which will inevitably increase the cost of monitoring and managing. With operations there is often a lower tolerance for risks because the impact can be substantial. The costs here represent the actions that we consciously take to try and prevent issues from arising (risks from triggering)—with our customer service example that might be training, coaching, *secret shopper* initiatives, audits, etc.

For all of our management costs we assume that the cost is for the reporting period—generally the fiscal year, and of course the template can be extended to add columns for actual and projected values to allow for the plan to be updated mid-year. It should also be noted that while these numbers are known, they are not fixed. Over time we may adjust and adapt our strategy, especially with something as dynamic as the project portfolio, and so these numbers still need to be regularly reviewed and updated to reflect the current situation in terms of risk management costs.

One final note on resource costs before we move on. We need to be careful not to *double count* the people that we assign to risk management activities. They are either an effort cost or a dollar cost, but not both. As a general rule employees will be considered as an effort cost, and contractors will be considered a dollar cost. It's perfectly acceptable to change that in your organization, but ensure that the guidelines are well documented and followed to retain consistency across the risks.

While all of the numbers that we have looked at so far are concerned with the funds we have committed to risk management activities, the risk capacity entry at the bottom of the table is a little different. Risk capacity represents the money and resources that we estimate we have available for risk management activities, and it probably requires some explanation as it is not a concept embraced by many organizations. If you think about a department within an organization, it has a known number of staff and a known number of positions (staff plus open vacancies). We can estimate the amount of work that has to be done within that department and determine in broad terms whether we have the right staffing level.

That's not so easy to achieve with risk management because it's not a department; it's a series of tasks that are distributed across the organization. We don't need an exact total here, so we can estimate based on things like the percentage of time that we expect an individual to spend on risk management activities. Financial numbers may be easier to predict based on previous years' financial statements, and a reasonable estimate is acceptable. For portfolio related activities, it's perfectly acceptable to have a set percentage of the effort and dollar budgets assigned to this category—we aren't looking for exact numbers but rather an indication of a potential problem.

Because these numbers are only estimates, we don't try and break them out by category but rather reflect them as a total for risk capacity at the bottom of each of the management columns. The intention is simply to provide an indication of the ability of the organization to fund the risk management activities to which it's committed without impacting its commitments elsewhere. It shouldn't be relied on as an absolute value (especially as it's a consolidated total for the organization) but rather as a flag that the organization is either stretched too thin in its ability to manage risks or that it is losing investment opportunities by reserving too many resources (dollars and/or people) for risk management that isn't required.

Understanding the Numbers—Risk Impact

Risk impact is slightly more complex because we have to consider the likelihood of the risk occurring. However, the calculation is still fairly straightforward—we take the risk impact and multiply it by the percentage chance of it occurring. If we look at the economic risk (recovery from recession slows) in our profile, the key numbers are:

- Chance to trigger (the chance to trigger [%] column)—20%
- Financial impact (Impact—Financial [$000s] column)—$2 million (number is expressed in $000s in the table)
- Resourcing impact (Impact—Resourcing [effort months] column)—100 effort months

We are looking to calculate two different numbers: the risk impact in financial terms and the risk impact in effort terms, and to obtain that we multiply each of the impacts by 20%. That gives us a financial impact of

$400,000 (20% of $2 million) and an effort impact of 20 effort months (20% of 100).

It is these numbers—the 400 and 20—that contribute to the totals, but they are not entered directly into the risk profile. Instead, they are used in calculating the numbers in the Exposure Totals row for financial and resourcing impact, respectively. These totals (the 2,560 and 80.6) are the sum of the impact multiplied by % chance to trigger for each risk, which is why the total isn't simply the sum of the individual entries (as is the case with management cost).

Okay, so the math is relatively straightforward, but the logic needs an explanation—why do we use 400 and 20? In reality the risk will either happen, in which case the financial impact is $2 million; or it won't, in which case the financial impact is $0. It therefore seems illogical that we use a risk exposure of $400,000 when we know that the exposure can never be that number.

The reason that it does in fact make sense is that we are looking at all of the risks that are faced by an organization, not individual risks in isolation. For example, suppose that we have five risks, each of which has a 20% chance of occurring. The law of averages says that one of those risks will trigger (because 20% = 1 in 5), so while the impact of that risk is more than the impact that we identified, that's offset by the impact of the risks that didn't trigger, which is less than we planned. In our example, if we had five risks, each with a potential financial exposure of $2 million and a 20% chance of occurring, then our calculated impact for each will be $400,000 (totalling $2 million); but only one of the five should trigger (20%), which will be a real world impact of the $2 million that we have set aside. Figure 5.1 shows this graphically and may assist in understanding; it can be a difficult concept to understand at first, but in the context of an organization that is exposed to numerous risks it does work, although we will consider some complications later in this chapter.

When we look at risks that have been identified from within the portfolio, we are distilling a complex set of risks into just a single line on the organizational risk profile. On each of our initiatives we will have conducted the same analysis that we have just looked at here—each of the risks will have been assessed a likelihood to trigger, a financial impact, and an effort impact. Then we will have multiplied those numbers together to come up with a total exposure for each risk and added those exposures together within each initiative. At the project/program/ portfolio level, this is the contingency reserve that we briefly looked at in Chapter 3 and will consider in more detail later.

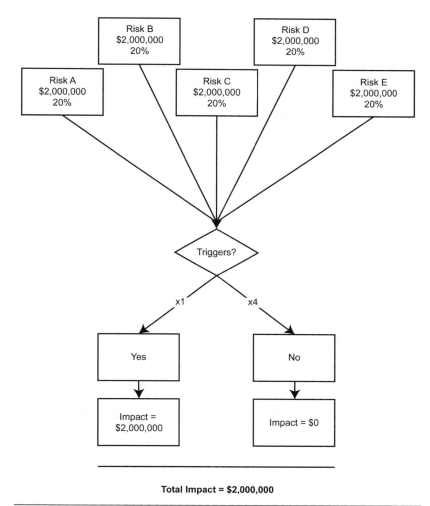

Figure 5.1   Risk impact calculation

However, when we transfer that lower level risk analysis from each portfolio element into the organizational risk profile, we take the total exposure—the exposure that we would be facing if every risk triggered, not the contingency reserve figures—and enter those numbers in the impact columns for each initiative. The reason that we ignore the project level calculations in the organizational risk profile is that we are going to enter a chance to trigger percentage at the organizational level, which will handle the conversion. This allows us to see the total potential exposure and avoids hiding potentially significant exposures simply because they have a low chance to trigger.

Technically the chance to trigger percentage that we enter for a program, or for a collection of portfolio projects, will be the weighted average of all of the risks that we considered in that program or project level risk analysis, but it's actually much easier to calculate than that. To calculate the percentage we just take the contingency reserve for the initiative that we are considering and divide it by the total exposure (we can use either financial or effort numbers, if you have done your calculations correctly it will give the same result). That will give you a number between 0 and 1 that you can easily convert to a percentage. When we then perform the calculations on the project and program entries in the organizational profile, the value that is entered in the impact columns will represent the contingency reserves for that project or program.

Now, all of this calculation of impact based on chance to trigger, and the use of contingency reserve numbers works perfectly if all of the risks are equal and if the law of averages plays fair, but life's not like that. If it is the major impact risks that trigger and the minor impact ones that don't, and/or a higher than average number of the risks trigger, then our exposure will be higher than we have identified in the profile and we are facing a problem. As a simple example, suppose that we have two risks, both with a 50% chance to trigger. One has a financial impact of $100,000 and the other has a financial impact of $2,000. Our contingency reserve will be calculated as $51,000 ($100,000 × 50% + $2,000 × 50%), but if it is the bigger risk that triggers, then we have $100,000 worth of impact and only $51,000 worth of reserves.

That's where the concept of management reserve becomes a factor, and we start to earn our money as experts in organizational risk. Management reserve is an important concept in risk management because it gives us an allowance for uncertainty. If we take a step back from the practical example of the risk profile for a moment, everything that we have looked at so far in this impact section has been about quantifying the unknown. We put a dollar and effort cost on risks and a percentage on the likelihood of the risk triggering. This is necessary to give us any hope of understanding and hence managing the risks that we face; we use our expertise to try and ensure that these numbers are as reasonable as possible. However, we have to be realistic here, *reasonable* is as good as it's going to get because risks are inherently about the unknown, and reasonable is not the same as *accurate*. Risks are unpredictable. They are driven by uncertainty, and the reality will not follow our nicely structured profile. That's why we need to allow for uncertainty.

Management reserve reflects that uncertainty. As we considered for contingency reserve, we have two values here, the financial and effort reserves. For some organizations, management reserve calculations are easy. They are just an arbitrary fixed percentage of the contingency reserve—sometimes a number as low as 5% (assuming that the organization acknowledges the need for a management reserve at all). A fixed percentage is better than nothing, but a custom approach based on exact needs is far better. If necessary, a compromise hybrid system that starts with a *base* percentage (20% is the lowest that I would be comfortable starting from) and then adjusts based on how the specific risk profile that you are dealing with varies from the norm.

There is likely to be a need for an increase in the management reserve if the following is true:

- There is considerable variation in chance to trigger and/or impact across the collection of risks, especially if high impact risks with relatively low chances to trigger exist.
- There is a lack of confidence in the accuracy of the contingency reserve calculations—perhaps because of a lot of unknowns or a lack of experience in dealing with the types of risks being faced.
- There is a lack of visibility into the risk—if there are no chances of *early warning signs* that the risk may be about to occur then there is no opportunity to prepare.
- Our risk analysis is incomplete or preliminary.

If the opposite of the above situations is true, then we can consider reducing the amount of management reserve.

Most of the above points are fairly obvious, but the first one probably needs more explanation. Take a look at the four different scenarios in Table 5.2. I have considered only the financial impact in those examples, but the concept applies equally well to effort impact.

You can see from those examples that the total contingency reserve for all four situations is identical—$60,000. However, the real risks that the organization might be facing are not at all the same. Suppose that risk A is the only one that triggers in each scenario. The actual impact on the project will be:

Scenario 1: $50,000, $10,000 less than our contingency reserve
Scenario 2: $50,000, $10,000 less than our contingency reserve
Scenario 3: $10,000, $50,000 less than our contingency reserve
Scenario 4: $250,000, $190,000 *more* than our contingency reserve

Table 5.2   Examples of how impact and chance to trigger drive management reserve

Scenarios / Risks		Chance to Trigger (%)	Financial Impact ($000s)	Contingency Reserve ($000s)
Scenario 1				
	Risk A	40	50	20
	Risk B	40	50	20
	Risk C	40	50	20
	Total			60
Scenario 2				
	Risk A	20	50	10
	Risk B	20	50	10
	Risk C	80	50	40
	Total			60
Scenario 3				
	Risk A	40	10	4
	Risk B	40	10	4
	Risk C	40	130	52
	Total			60
Scenario 4				
	Risk A	20	250	50
	Risk B	50	10	5
	Risk C	50	10	5
	Total			60

It's pretty obvious that we have a serious problem with scenario 4, but what caused it? The issue is that the chance of the risk occurring is relatively small, just 20%, and so the contribution to the contingency reserve is also fairly small—$50,000. That is fine if all of the risks have a similar impact (as in Scenario 1), but because the other risks in Scenario 4 have minimal impact then even though they have a higher likelihood to occur, they aren't able to contribute anywhere near enough to the reserve to cover the shortfall from Risk A.

The opposite situation occurs in Scenario 3. Risk A is the low impact risk, and so we come in far below the contingency reserve, because Risk C made a large contribution that ended up not being needed. Scenario

2 demonstrates that variations in chance to trigger alone do not materially impact things, as long as the overall risk remains consistent. While organizations might like to think that Scenario 3 is what will happen to them, hopes and dreams don't make for effective risk management. The executive who signed off on an arbitrary 5% of contingency reserve as the management reserve figure would have some difficult questions to answer in Scenario 4.

It should be noted that in both scenarios 3 and 4 the management reserve should be *increased* because of the risk of having high impact outlier risks. The only difference between scenarios 3 and 4 is the risk that triggered. If Risk C had triggered in Scenario 3, we would be in just as much difficulty as with Risk A triggering in Scenario 4.

Determining the provision for management reserve requires a strong understanding of the risks that the organization faces, and while it is only expressed as a single number in the management reserve row of the organizational risk profile ($500,000 and 24 effort months in Table 5.1), there will actually need to be a lot of thought and consideration given to it. For projects and programs within the portfolio, the analysis is done for each initiative and can roll up to the organizational level as an addition to the total. However, for other risks, especially external ones, it will involve a combination of experience, understanding, and good judgment that will require input from the various risk owners. This is the one area of risk management where the best processes, the most accurate analysis, and the best sources of information simply cannot be enough. We need to rely on the ability of risk experts to interpret the organizational risk environment and make appropriate determinations of the required management reserves.

Any statisticians reading this may be tempted to take a more scientific approach using standard deviations of the individual risks, but in the real world, that's not really going to help; risks are still uncertain and unpredictable.

One final note on the reserves that we have discussed in this section: the contingency and management reserves represent costs that the organization is expected to incur. Although the expectation is that there will be an allocation of financial budget and an under allocation of resources in order to cover these costs, that allocation still has to take place, and that's what we'll look at in the next section.

Understanding the Numbers—Capacity

So far we have focused on the cost side of the ledger, the costs that we know we will be incurring through our attempts to manage risks and the costs we anticipate will affect us as a result of risks triggering. Now it's time to try and *balance the books* by assigning financial and people resources to cover these costs. Let's start by looking at the type of funding we need to provide:

- *Management cost—financial.* This funding is the money that we budget to cover the costs that we have accepted for risk management. It might be insurance premiums, outsourcing costs, even things like costs to upgrade buildings to make them more hurricane resistant.

- *Management cost—effort.* Here we consider the resources assigned to risk management activities—the tasks that we carry out on projects, the time spent looking for ways to improve the organization's ability to withstand a downturn in the economy, etc.

- *Impact cost—financial.* There are a number of different items that make up this category. First there is the money that we put aside to deal with triggered risks—the money that we spend on implementing contingency plans. Second, there is the money built into our budgets for lost revenue—a safety margin between expected revenue flow and budgeted revenue flow to deal with risk events that reduce or delay that revenue. Finally, there is the allowance in the budget for reduced savings—to allow for cost reductions that are lower than planned due to triggered risks.

- *Impact cost—effort.* This category represents the allowance that has been made for the assignment of resources to deal with triggered risks—to implement contingency and facilitate recovery. Most of this cost is related to the work necessary to implement contingency plans, but this will also reflect the effort needed to deal with the consequences of delayed savings—the *safety margin* in staffing that we maintain between what we need and what we have.

As you can see, the capacity is not simply the sum of the costs—it is the money and effort that the organization specifically assigns to offset those costs. If the risk costs are fully funded, then the two will balance (we already allow for uncertainty in the profile with the management reserve).

For the funding assigned to cover the consciously accepted costs of risk management activities, the total is calculated in a similar way to the costs. It's the sum of the budget items assigned to the costs, which in most cases will be the same as the planned costs themselves, with perhaps a small safety margin to deal with changes in plans. Think of a project as an example; we identify the tasks to be completed in our project plan, and a subset of those tasks will be related to risk management. If that plan is approved, then effectively we are issued the money and effort budget to cover the costs.

For external risks there may be a more arbitrary approach—a percentage of a role as defined by a job description for example, and that's generally fine as long as the job description has been maintained and accurately reflects the work that is expected of the person in the role.

When we consider risk impact, the funding is based much more heavily on judgment by the organization's leadership—executives decide whether to fully fund the identified risk impacts, whether to overfund, or whether to consciously maintain a shortfall. For funding of the financial elements of the anticipated risk impact, there are three tools that the organization's leadership has available:

1. Direct funding—we can keep back a portion of the annual budget that is available for investment to deal with the impact of risks that become real. This has the advantage of being readily available to the organization if it is needed and can be released as soon as it is needed—it's real money (or at least real available credit). However, there is a cost to maintaining these funds. If they were used for other purposes, then they could potentially generate additional gains for the organization. That might mean that the organization could fund another project, provide additional resources to existing initiatives or operational areas, or even invest more heavily in risk management.

2. Revenue loss—we can adjust our revenue projections downward if we think that triggered risks will reduce revenue (or slow revenue growth). This is an indirect funding model for risks in that it does not directly require funds to be assigned today but rather uses future revenue to pay for the exposure. Essentially this approach is to take the expected benefit from delivering the revenue generating projects and then reduce it by the affect that triggered risks are expected to have on the actual project deliveries. This is a tempting way to *pay* for risks because the money isn't seen as being as real, but it can only apply to risks that have the potential to impact revenue, and it can have a far reaching impact—revenue lost to a competitor today may not be regained for many years, if at all. In many organizations, this approach is used heavily but it is hidden— the goals that product owners sign up for are already adjusted

down to allow for these risks, effectively hiding the potential revenue that is being lost to risk events.

3. Operational cost—we can adjust our operational cost projections upward if we believe that risk events will drive additional costs into the organization, or if they will delay the ability to realize operational savings. Like revenue loss, this is an indirect funding model in that it's paying for the risk exposure with future costs. Again, this can only apply when the risk triggers have the potential to drive costs into the business although these are less wide reaching than revenue loss—we can usually drive costs out of the business relatively quickly once the risk exposure has been removed. This is another commonly used approach within organizations and again tends to be hidden within operational departments before the goals are established, hiding the operational cost impact of triggered risks in the process.

In reality an organization will likely need to use all three of the above approaches to *pay* for risk exposure, although care should be taken relying on revenue loss and operational costs to cover large percentages of the exposure—that can rapidly drive the wrong behavior into the organization. If people feel that the budget is already allowing for the costs of the risk impact in the revenue and cost numbers, then there can be less incentive to manage risks effectively, which drives more risk exposure into the organization and increases the need for risk capacity.

Effective management of these categories of capacity not only ensures that the risk exposure is covered, it can also help to ensure that the organizational priorities are preserved. For example, if the organization is focused on gaining market share and growing revenue, then funding of risk exposure should come from a larger amount of the budget being put aside for risk recovery (direct funding) and from a greater willingness to absorb additional operational costs in order to minimize the impact on the revenue that is the corporate priority.

It should be clear by this point that while we have split risk impact and risk capacity in our analysis here, the two aspects need to be developed in parallel to ensure that the appropriate risk recovery (and management) strategies are developed—we can't consider them in isolation because they are inextricably linked. We can't develop effective risk management plans without understanding roughly how much capacity is available for management, and we can't develop a budget to pay for risks without some understanding of how much exposure the organization is facing.

One final note on capacity—we have talked here in terms of funding as if the organization has a choice, but I have called the category *capacity* for a reason. Many organizations may simply not have the funds or resources to cover the total risk exposure plus the cost of risk management activities. If that is the case, then the organization should look for ways to reduce the overall risk exposure to a level that it can fund. In reality, the organization will often go ahead and find itself in a situation that it cannot recover from without significant impact on costs, revenue, customers, etc. If the organization consciously chooses to underfund risks, then it needs to understand that it is leaving itself exposed to some serious consequences—that may be acceptable in some circumstances, but the worst case scenario has to be considered and accepted as a possibility.

Analyzing the Profile

We've spent a lot of time so far in this chapter looking at the individual elements of the risk profile in a lot of depth, and now it's time to pull the whole thing together and look at what the profile is telling us. It's important to understand that the organizational risk profile is only a summary of what is happening within the organization. Any decisions we make to change our risk management strategy should be based on the underlying risk analyses that have rolled up to this profile in order to ensure the actions we take can be implemented within the business. This profile merely helps to identify where that drilling down may need to occur and assists us in seeing the overall risk situation for the entire organization.

It's also important to understand that everything within the risk profile is connected. If we choose to reduce funding on the management side, then we likely have to reduce the costs that we are capable of assuming to manage the risks and that will likely drive an increase in risk exposure, which in turn will require an increase in funding of contingency and management reserves. If we want to consciously drive costs out of the profile, then it likely requires a change in the operation of the business—often a change in the initiatives that are undertaken as part of the portfolio.

So let's look at Table 5.1 and see what we can determine about the hypothetical business from it. Obviously a real risk profile will be far more complex than this, but this will help us to understand the principles that are at play and learn how to interpret the profile.

First of all let's look at the individual risks and see what that tells us:

- For our external risks there is minimal active management occurring, suggesting that there is little that we can do to influence them, and that makes sense—the action of competitors or the economy is largely beyond our control, so it would be more concerning if there was a lot of management effort being directed there.

- The external risk that does have management costs against it also makes sense—some lobbying effort or political contributions to try and influence that. The likelihood of that risk occurring is still fairly high, so we may want to consider whether more money and/or effort should be devoted to that. However, the impact is relatively small—$250,000 so perhaps the additional effort wouldn't be worth it. Lobbyists are expensive and can only have limited effectiveness. Additionally there are limits on the amount of political donations that can be made. We may even consider reducing the management there, but we need to understand what that will do to the exposure.

- There are two external risks that have large impacts if they occur—$1.25 million for one and $2 million for another, and yet with low likelihoods of triggering (10% and 20% respectively). As we saw above, that's an indicator that should lead to an increased management reserve. However, the management reserve is still fairly low—less than 20% of the total exposure, so that may be a cause for concern given the size of the potential exposure from those two risks.

- Risk management of internal risks, especially for the three items in the financial category (which is where our projects will generally sit in this example, as we don't have strategic or technological in our profile) seems to be skewed toward financial cost rather than effort cost. That will have rolled up from the individual projects so may well be correct, but it will be worth checking that the approach is consistent with the portfolio constraints. It's not a dramatic shift toward dollars over effort, but it is there—only 15 effort months (4 + 5 + 6) dedicated to managing financial risks but $200,000 ($50,000 + $70,000 + $80,000).

- The impact and likelihood of occurrence of some of the financial risks are still quite high, and we would want to satisfy ourselves that we couldn't reduce those further with the acceptance of a little more management cost. Again, the analysis will have been conducted at the project and program levels, but it will be worth validating that there has not been a focus on reducing management costs based on project constraints to the detriment of effective risk control.

- The customer service risk management costs seem remarkably low for the amount of exposure that remains, and intuitively we would expect customer service to be fairly responsive to risk management activities—training, coaching, process, and policy enforcement, etc., should help to maintain standards, which in turn will reduce the instances where customer service fails. This is backed up by the high potential impact—$1 million dollars. While the chance to trigger is relatively low (another indicator that our management reserve may not be high enough), this $1 million dollar potential exposure is way too high and should set off warning bells that we need to revisit this risk area with a view to increasing the management cost and reducing the exposure. If this can't be done within the current environment, then this may be an early warning sign that the customer service processes are broken—operational processes should be responsive to risk management.

Before we move on to look at the total exposure and capacity, let's just consider the points above. From a fairly basic assessment of a few risks we have identified a number of areas where we should be digging deeper to understand what is happening in the organization. This is powerful—this risk profile may require effort and commitment to develop and implement, but it can identify real issues in the organization early enough to be able to make a significant proactive impact. The customer service example is a good one. The profile makes the problem evident—we are spending only $40,000 and the equivalent of half of a full-time resource to try and control something that could cost us $1 million (and that doesn't consider the reputational risk). That's a dangerous strategy and one that can have long-term impact—it takes a long time to recover from a reputation for bad customer service. If this were my organization, I would be conducting an immediate analysis of what is going on in this part of the business.

Now, let's move on to look at the summary rows at the bottom of the profile and see what they tell us. The financial management reserve, at around 20% ($500,000 management reserve divided by $2,560,000 total financial risk exposure), may look okay at first glance, but we have identified a few areas above that are likely to drive a need for an increase, potentially a significant one. Virtually all of our contingency reserve of $2.06 million (the total financial exposure less the financial management reserve) could be wiped out if the economic risk triggers. We really should be looking to dramatically increase the management reserve unless we are being deliberately conservative with our risk exposure in other areas of the organization because of the apparent economic uncertainty, and that doesn't appear to be the case. We may think that the resourcing management reserve is in better shape because that number is nearly 30% (24 divided by 80.6), but in some ways this situation is worse. The economic risk impact is 100 effort months, so if that risk alone triggers, then we are considerably over our total calculated exposure of 80.6 effort months.

The total risk management cost seems to be low, less than the equivalent of two full-time employees managing all of the total risk exposure, and only $260,000. However, remember that a number of the external risks can't be managed, which will reduce the numbers, although we have identified a potential problem with the customer service risk management that will need an increase. The capacity for management costs is only just adequate—especially on the effort side—and that may be a source of the low costs. The organization may be resisting spending more money on risk management and that is restricting the ability to manage risks.

The surplus or shortfall values are the most immediately critical when we consider the totals. It's obvious to say that we should ensure we have provided sufficient risk funding to cover the exposure, but it's more complex than that. In this particular example, it would appear as though we have a shortfall in the financial exposure capacity—we may not have enough money put aside to cover the exposure. Given that we identified above the potential need to increase the management reserve further, the shortfall could become even more pronounced than the $60,000 that we see here. At the same time, the financial impact is 97.7% funded (2,500 divided by 2,560), so we don't have a significant shortfall, and the organization may consciously decide to accept that shortfall in order to avoid the alternative. Remember that there is a cost associated with providing capacity (we either have to reduce investment elsewhere in order to keep cash reserves, reduce revenue projections, increase cost projections, or some combination thereof).

In this case the organization may have consciously decided to carry the shortfall and make adjustments elsewhere during the year if necessary—cancel a project, impose cost cutting measures (reduced travel, hiring freeze), or look to improve revenue (anything from a sales spiff, to a product promotion, to a price adjustment). If we start to see capacity at less than 90% of exposure, then we have a greater reason to worry, because we are getting to the point where we can't realistically recover and may be facing problems—in large companies, this number may move higher to 95% or even more as 1% can represent significant amounts of money (or effort). In this situation, a more realistic management reserve will likely drop the coverage below 90% and indicate that an increase in funding is required.

My bigger concern with our risk profile as it stands is actually with the one that looks to be in the best position at first glance. The financial management capacity of $300,000 is quite a lot higher than the actual exposure at 115.4% of the costs (300 divided by 260). Forget the fact that the numbers are small in our example (they could be millions instead of thousands); we are assigning over 15% more capacity than we actually require to manage the risks effectively. As indicated above, the excess capacity (and more) might actually be required based on the adjustments that we talked about making to the customer service risk category and to risk management as a whole, but at the moment that represents a bad use of funds.

While we may never end up physically spending the cash, we are tying up money in this budget item that could be used elsewhere to help the business achieve its goals, and that's not a good thing. If we are still working with preliminary estimates that may change, I can accept an over-allocation in that case, but as we move through planning and our numbers become firm, we should be looking at no more than 105% of the cost for our capacity on the management side.

One data element that will be relevant to the total costs and total capacity is the comparison of management and impact totals over time. We have seen that this snapshot gives us a lot of information and identifies areas where more work is needed. However, when we add in the ability to look at trends then we add another element.

Changes from year to year may be dramatic depending on the make-up of the portfolio, the impact of events that occurred last year, or the changes in strategy, but we should be able to see alignment with our corporate priorities. In shorter time frames (and this should be a living document that is actively managed and maintained), we would expect to see indications of whether we have made the right decisions around risk management. If the organization utilizes an annual planning cycle, then we should see risk exposure and risk management costs decrease as initiatives are completed, but we will also see a reduction in capacity as triggered risks eat into the money put aside to cover contingency and management reserves. In particular, we should be looking for signs that capacity is dropping faster than exposure, as that could be an indication of serious problems ahead.

Over longer periods, if we are becoming more risk tolerant as an organization, then we would expect to see increases in risk exposure. If we are becoming increasingly risk averse, then we would expect to see more money spent on risk management, even if the incremental impact of that additional spend is reduced.

That's been a lot of analysis to get to our organizational risk profile, and we have gone to a lot of depth, but this tool is the basis of understanding and controlling organizational risk. Everything that follows builds on this concept and helps to connect the activities within the organization with their impact on this profile.

Ownership of the Organizational Risk Profile

Ensuring that the right ownership model exists for the risk profile is vital to success. Each risk category will have an owner, but someone has to drive the profile itself. I strongly recommend that ownership and accountability reside with the head of the organization's EPMO or with the portfolio management function. If the organization has a Chief Project Officer then that is also an appropriate owner. If the organization has multiple PMOs based on department, then I have no problem with separate profiles being developed for each PMO's initiatives, but these ultimately have to be consolidated to one profile for the organization as a whole with a clearly defined owner.

While not all of these risk categories are directly related to the project portfolio, those will be the most dynamic risks and likely the ones with the highest amount of management activity. Additionally, risk management expertise lies within the PMO/EPMO/portfolio management functions, and so the analysis will be easier if ownership resides there.

However, the final accountability for ensuring that the organization has a risk profile and uses it to drive risk command and control activities must lie with the executive level of the organization, and ultimately the CEO. If the organization does not support the concept then there will never be an effective profile, and by extension, organization risk command and control will be disjointed or non-existent.

This book has free material available for download from the
Web Added Value™ resource center at *www.jrosspub.com*

SECTION 2

6

The Risk Management Partnership

In Section 1 we looked at risk categories, the relationships between risks, the impact that risks can have, the variables that contribute to a command and control structure, and finally, at the concepts that result in an organizational risk profile. In this section, we'll look at those same concepts from a practical standpoint—we'll build and execute a process framework that supports command and control in order to manage the relationships and impact within the organizational risk profile.

Inevitably, we will look at organizational risk management from the perspective of different functions within the project execution organization—most notably portfolio management, program management, project management, and the PMO. It's important to understand that we are talking about a single cohesive approach, not a number of different ones for each of the functions. We are going to be focused in this section on the organizational processes, but the process framework is deliberately structured to have commonalities with typical risk management processes. We can't view portfolio or program level risk management as being fundamentally different to project level risk management; rather, there is a relationship between the different levels—a partnership that aligns all forms of risk management to maximize the chances for risk management success at all levels.

There are two main elements to that partnership:

1. The processes that are applied at each level of the organization should ideally leverage one another to facilitate effective risk management. At a minimum, there should be safeguards to

ensure that there is no duplication of effort and that there are no gaps between the processes at different levels where things can get missed.

2. The people who execute risk management at each level of the organization. There should be clear and complete communications between the different groups at the portfolio, program, and project levels to ensure the actions taken are consistent and complementary, helping to create synergy rather than working against one another.

Let's look at each of those two elements in a little more detail.

Process Partnership

Clearly the portfolio, programs, and projects are closely connected, and from a management standpoint the alignment is almost total. A portfolio manager may have responsibility for some new project candidates that are in development, and for ensuring the benefits are achieved for some completed initiatives, but the vast majority of his or her work is concerned with the execution of the current portfolio of projects. That means managing work that is being undertaken at the project level, a situation mirrored on a smaller scale for program managers.

Given that the projects are the execution vehicles for portfolio and program decisions, it shouldn't come as a surprise that there is alignment in the various management processes. Organizational risk management is a more strategic activity than project level risk management and has to consider more variables, but the essential actions that are taken will be very similar to project risk management, something that we will see as we develop our risk management process in the next few chapters.

This has the advantage of making it much easier for project level resources to assist in the execution of organizational risk management activities because they will recognize most of the actions that they are being asked to take. Consider a situation where a front-line project resource is suddenly asked to contribute to the management of a portfolio risk without any prior warning—something that could easily occur. If the mechanics of the work that the project team member has to execute are familiar from work they have previously conducted for project level risks, then they will find it much easier to be able to focus on the specifics of the risk. If the process itself is different and unfamiliar, then some of their focus will be diverted to understanding the process. Their ability to effectively monitor and manage the risk itself will be reduced.

Further, if processes are well aligned, that same resource can conduct portfolio, program, and project level risk management in one combined set of activities. The specific management steps undertaken to manage a risk may change if a risk moves from being only the subject of project level focus to being managed for the potential program and/or portfolio impact. However, the overall management framework should be able to remain the same—progress reporting to the portfolio, program, and project should be able to be combined. This may not happen when processes are first deployed, but it should be the goal as processes evolve and become more widely accepted within the organization.

That doesn't mean that the management of the risk is identical at each of the different project execution levels; there may still be a number of key differences that have to be considered:

- Aggressiveness of management. Because the portfolio, program, and project each has a different perspective of the risk and how it may affect them, there may be a different view on how aggressively the risk needs to be managed. A simple example might be a risk that has the potential to cause much more damage to a project than to a much larger portfolio, but this may not always be the case. There may be situations where project resources carry out much more aggressive risk management because of the potential damage that can be done higher up in the organization.

- Management approach. There are a couple of different ways that the approach to risk management can be affected by different levels of the organization. The first is similar to the aggressiveness of management discussed above—the focus on the risk at a higher level may result in adjustments to the way the risk is managed. The second reason the management approach may be changed is more related to the organizational risk profile that we developed in the previous chapter. Portfolio management is responsible for managing all elements of risk within the portfolio. That will also mean managing the overall investment in risk management—driving costs between effort-based measures (mitigation for example) and dollar-based measures (typically transference), or modifying the overall risk tolerance based on the constraints hierarchy, which we will explore later in the book.

Process is only half of the story, and we will look at the people side next, but if organizational and project risk management processes are aligned, then the ability to integrate the additional complexities of portfolio and program level risk management becomes easier, if not seamless. As you continue reading through the rest of this section, and particularly as you look at Section 3 on process implementation, consider organizational risk management as a process that needs to align and integrate with existing risk management approaches. You should never compromise the integrity of organizational risk management for the sake of alignment, but you should also never pass up opportunities to align with existing processes where possible.

People Partnership

Risk management is an unusual discipline in that it requires some specialist skills, and yet it has elements that are executed by people who are not in specialist roles. Project level risk owners are not selected because of their skills and experience at managing risks but because they are the most appropriate person to monitor and manage the risk specifics—the trigger events, early warning signs, responsiveness to management actions, etc. That generally works within a project because team members have a reasonable understanding of what the initiative is trying to achieve simply from being a part of it and can therefore judge when things are starting to go wrong.

However, when we create separation between the person monitoring the risk and the impact that the risk can have, as is the case when project resources are managing risks that have program and portfolio level impact, then we start to create uncertainty among resources:

- *Risk owners.* The people who are managing the risk at the front line likely no longer feel that they fully understand the significance of the risks that they are responsible for because they don't have visibility beyond the project.
- *Portfolio and program managers.* By handing off risk management to people who are a relatively long way removed from the strategic drivers that will be impacted by the risks if they trigger, there is concern about an increased opportunity for mistakes to occur.
- *Project managers.* The connection between portfolio and/or program management and the project level resources who own the risks can leave project managers feeling excluded and out of touch with what is happening on their initiatives.

All of these concerns are understandable, but none of them is acceptable if organizational risk management is going to be effective. Fortunately, these uncertainties should be easy to address through effective, proactive communication, and this is where the concept of a risk management partnership really comes into its own. Let's consider a very simple, single example of a portfolio level risk that is likely to display early warning signs on a single project within a program. To effectively manage this risk, we need engagement at the portfolio management function that is responsible for identifying and analyzing the risk and the project team member who will be watching for signs of trouble as part of their work. We also need to ensure that the program and project managers are aware and engaged so that they can manage their initiatives with an understanding of the needs of the portfolio. If communication breaks down at any of these points, then not only do we have less effective risk management, we have the potential for relationships to be damaged if people feel consciously excluded.

There are legitimate issues when the execution of portfolio level risk management actions rely on project level resources who don't have the same understanding of the strategic environment as portfolio level resources. However, in the vast majority of organizations, there is no alternative; portfolio and program management levels are not heavily staffed and may consist of just a single resource (and at the program level, a program manager may have responsibility for more than one program). This is again where the concept of partnership has to come into play. A hierarchical structure that requires information to be passed through multiple levels until it reaches an expert for assessment and interpretation and then requires the decision that comes from that interpretation to be communicated back through those same levels to the person who needs to implement the decision simply won't deliver effective risk management.

Another scenario where a people partnership is important in organizational risk management is where specialists are needed—people within the organization who have specific skills that are required in order to effectively perform certain elements of risk management. One of the easiest examples to consider is risk analysis. Risk analysis requires a detailed understanding of the specifics of a risk situation, the portfolio environment, the risk tolerance, and the connections between different elements. That's way too much for one person to control and requires a number of subject matter experts to be engaged. These experts may specialize in risk analysis itself—perhaps someone from the PMO—or they may specialize in a particular situation that only impacts one or two risks. Regardless of the situation, their engagement in organizational risk management is limited—either to a subset of the processes or to a subset of the risks. For risk management to be effective, these specialists need to be able to smoothly and efficiently move into their role, perform the required tasks, and move out again without causing disruption. While some of this is down to process, a lot of it is down to a people environment that is flexible enough to accommodate the needs of these specialist roles without causing disruption. When you rely on expertise, there is frequently a need to consume that expertise at the convenience of the expert, and the people need to act as the buffer between the fairly rigid (deliberately) processes and the rigid schedules and availability of the experts. This may not be what we typically think of as partnership in the same way as resources from the different portfolio levels will partner together, but it is just as significant. Remember that this is risk management that considers the whole organization (and beyond) within scope.

The final point to make with regard to people is that as well as forming a partnership with one another within the portfolio and the organization beyond the portfolio, they also form a partnership with the processes. This is logical when you think about it—processes are executed by people, and people are guided in the way that they operate by the process framework of the organization. This needs to be an intelligent partnership with the processes guiding the actions of people rather than dictating them—a set of recommendations rather than rules if you will.

This partnership is vital to consider as we develop and implement our organizational risk management framework in the rest of this book. If process dictates the actions of people, then we lose the expertise and judgment that we invest in so heavily in our organizational leaders, and if people ignore process, then we lose all benefits from the organizational risk management concept.

Beyond Risk

This book is about risk and that's where we are going to focus, but the concepts we are considering here can be applied equally well to any other element of portfolio execution. If you successfully implement portfolio risk management within your organization, you aren't then *done*, it's simply time to determine what to do next—an organizational quality management process perhaps? It's difficult to consider the concept of these different process elements working in partnership if risk management is the first process that you develop, but you will likely still have some elements of legacy portfolio management process to integrate with your new risk processes.

Even if these processes are nothing more than high-level portfolio tracking and reporting processes that are simply an administrative overlay to your organization's project management methodology, you will need to ensure your risk management process can work with those elements rather than against them. The strength of the portfolio management approach will help to determine the dominant player in the partnership. If you have well-developed, broadly accepted, and effective portfolio management in place, then risk management needs to be able to integrate with that approach with a minimum of disruption. The risk management processes should never be compromised by the portfolio management approach that exists, but it should attempt to align with it wherever possible.

On the other hand, if your portfolio management approach is underdeveloped and inconsistently applied, then you should be prepared to drive change into it with the development of your risk management processes. For example, if effective risk management demands access to information at the portfolio level that is not currently available, changes may need to be made to the overall portfolio management and control approach in order to make that information available. That may well result in an organizational risk management implementation driving additional work into the organization, and that's okay.

Organizational Partnership

The very last thing to consider before we start developing our processes is the partnership with the organization as a whole, and in particular with the executives who drive organizational behavior. This partnership needs to be a little more formal than most of the others, and it is vital to success—it will ultimately determine whether organizational risk management can deliver lasting benefits. Effectively this partnership formalizes the commitments that are involved in the decision to develop and implement organizational risk management:

- Commitment to the process. Recognition that an organizational risk management process delivers benefit to the organization and that once an effective approach is developed and implemented, it will be utilized throughout the portfolio to improve the quality of portfolio execution.
- Commitment to support functions. The most obvious of these is the development and maintenance of an organizational risk

profile as we looked at in the previous chapter. This is a foundation document for understanding the organization's risk environment, which in turn drives the risk management approach. There will be less obvious elements—processes to generate process inputs, increased frequency of existing processes, additional detail in other areas, etc.

- Commitment to continued support and investment. Any new process will take a while to become embedded and accepted, and it will need to evolve as that acceptance progresses. The organization needs to recognize that and commit to staying the course—not blindly funding a failing process indefinitely but realizing that the payback period is going to require a period of several years, especially in a relatively slowly repeated cycle (such as with an annual planning process).
- Visibility of commitment. Different elements of the business will be at different degrees of readiness for an organizational approach, and as a result, some business leaders will take longer to appreciate and embrace the benefits that the processes will deliver. However, there needs to be a commitment to visibly supporting the development and implementation rather than being apathetic or even negative toward it.

This may sound a little bit like a contractual relationship, and in some ways, it is. It shouldn't be viewed as confrontational negotiation, but it is important to ensure that the organization is making a commitment to the approach in full understanding of what is involved. This needs to be viewed as a partnership that will develop and grow over a period of years, and it is a commitment to a fundamental shift in the way that a core corporate function—the achievement of goals and objectives through project execution—is achieved.

And that seems like an appropriate point to start that shift and begin to develop our processes.

7

The Organizational Risk Management Process

Later in this section we will look at the specific considerations of risk management at each of the different levels of the organization—portfolio, program, project, and PMO—but first we need to develop a process framework that we can apply to each of those elements of project execution. There will be variations on the approach at each level, and we'll consider those in more depth when we look in detail at portfolio and program level risk management in particular. However, before we can start worrying about exceptions, we need to build a framework that is capable of supporting strategic risk management for the organization as a whole. It should be noted that while the focus of this book is on the application of organizational risk management within a project portfolio, the same process framework works just as well for the external risks that we looked at in the first part of the book. The processes can be lifted out of a portfolio setting and inserted into business unit operational processes. This will not only provide further consistency in the approach (see the discussion in the previous chapter on process partnership), it will also help with the alignment of organizational risks that trigger and impact the portfolio.

Let's begin by considering one of the major drivers of our behavior in organizational risk management—where risk sits relative to other constraints. This is vital because it helps us to determine how important risk is within the context of the portfolio and what actions are available to us (in a general sense) to manage those risks. To do that we look at a tool called the constraints hierarchy.

The Constraints Hierarchy

Before we can start managing risk (or any other aspect of the portfolio) effectively, we need to understand what the organizational attitude toward the portfolio constraints are. I grew up in project management with the concept of the *triple constraint*—budget, schedule, and scope. The principle was that if you changed one of those constraints, then you had to change at least one of the others to maintain a balance, otherwise you didn't maintain the triangle of those constraints—the *triangle diagram* that is shown as Figure 7.1.

Well today, PMI has expanded the concept of the triple constraint to something that I believe is a more appropriate consideration of six different constraints. They have split cost or budget into two separate items—dollars and resources—and they have also added quality and risk to the mix, resulting in the following six constraints:

1. *Budget.* The money that has been formally made available for the initiative.
2. *Quality.* The required standard that the project has to meet. This may be measured in many different ways depending on the type of project—number of failed items per 1,000 in a production environment, number of bugs in a software project, etc.
3. *Resources.* The nonfinancial elements that will be used/consumed on the project. The obvious consideration here is people and effort, but it may also include machinery and equipment, office space, and so on.
4. *Risk.* The level of risk that the initiative can accept, usually measured in terms of exposure (amount of required reserves) for dollar and effort impact.
5. *Schedule.* The amount of time assigned for completion of the project, usually expressed in terms of the date by which the work has to be completed.
6. *Scope.* The work that is expected to be completed by the project and/or the deliverables that will form the output of the initiative.

Obviously, by definition the constraints are limiting factors on the projects, programs, and portfolio, but those constraints are also fundamental to understanding how the portfolio and its constituent elements need to be managed. As risk is one of the constraints, it is important to understand how the constraints drive portfolio management before we start to try and manage risk. The constraints hierarchy concept can be applied

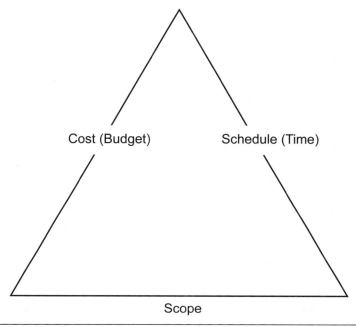

Figure 7.1 The triple constraint

at all levels of the portfolio, but we should be concerned predominantly with the portfolio level, as that is the overarching element. Program and project level constraint hierarchies can drive behavior, but only within the portfolio framework.

The constraints hierarchy is simply a sequencing of the six constraints from most important to least important. At the project and program level, this is agreed to by the stakeholders and documented as part of the charter. However, at the organizational level it represents more of a set of guiding principles that assists us in making decisions that impact the portfolio and portfolio elements. Let's start by looking at the concept and then apply it to the organization.

The constraints hierarchy is established by first establishing which of the six constraints must be preserved at all costs. In some initiatives, it's easy—in a major sporting event like the Olympics, the date has to be the most important constraint. You can't ask the athletes to come back in a few months because the stadiums aren't quite ready; you just have to do whatever it takes to get the work completed on time. If that means sacrificing the dollar cost, the number of resources, some of the nice-to-have features, the quality of noncritical elements, and the organizational risk exposure, then so be it. No matter how unpalatable the decision, the date has to be preserved.

Not all projects have such an obvious top constraint, and the process for determining the most critical constraint can be complex. Even when it is simply a stakeholder-driven decision there can be significant disagreement between different stakeholders, and the portfolio level should provide some guidance in the case of disagreement, but we need to establish the most important constraint at the portfolio level first. If the most important portfolio constraint is not obvious, then we should be looking for guidance in the organization's goals and objectives.

For example, suppose that the objectives for portfolio are focused around operational efficiency and cost saving. That would suggest that the organization will want to have budget and resources as some of the most important constraints—it doesn't make sense to be trying to save money and not be concerned about the cost of achieving those cost savings. Conversely, if the organization is in expansion mode and is looking to aggressively increase market share and revenue, then schedule and scope are likely to be important. The stakeholders—often the executives at the portfolio level—will still need to be engaged in confirming the most important constraint, but if that decision is inconsistent with what the organizational priorities are indicating, then we may have a fundamental problem in the structure of our portfolio!

Once we have established the constraint that goes at the top of the hierarchy, we repeat the process to establish the next most important constraint—the last of the remaining five constraints that we will sacrifice, or put another way, the constraint that we will only sacrifice in order to preserve the most important constraint, not for any other reason. In our Olympics example, quality might be appropriate here—we'll sacrifice quality if there is no other choice in order to meet the schedule deadline, but only if every other option has been tried first and quality will only be sacrificed to save the schedule, not for any other reason.

We continue this process for the remaining constraints until we have a full hierarchy of the six constraints with the least important (the one that we are most prepared to compromise) at the bottom, and the most critical at the top. Of course, this doesn't mean that the lowest of the constraints in the hierarchy is not important; we will still do whatever we can to protect it with changes that don't impact our constraints, but it identifies the first place that we will look to make compromises if it proves impossible to protect all of the constraints.

From a risk management perspective the constraints hierarchy guides us in the activities that we can undertake in order to control risks. If risk is the most important constraint (at the top of the hierarchy), then we have maximum flexibility to manage risks—potentially any action can be taken that we consider necessary. On the other hand, if risk is the least important of the constraints (at the bottom of the hierarchy), then we are much more limited in what we can do.

These are still only guidelines, not hard and fast rules—we wouldn't refuse to spend $1,000 to eliminate a critical risk just because budget was higher than risk in the hierarchy, and we wouldn't delay a project by a month simply to make a fractional reduction in risk exposure simply because risk was at the top of the hierarchy.

Sequencing of Organizational Risk Management

In a project management environment, the risk management activities are fairly straightforward. At the start of the project, we conduct our risk identification and analysis; we then prioritize the risks and develop management approaches and contingency plans. Once that has been completed, we begin the execution of risk management. Throughout the project, we review the progress and the status of risks, making adjustments as necessary.

At the organizational level, things aren't that straightforward. Even if the organization operates on an annual planning cycle that sees most initiatives wrap up by the end of the fiscal period, the portfolio itself cannot operate on the same series of initiate/plan/execute/close cycles—there will always be new project candidates that are being reviewed and completed projects where benefits realization management is being undertaken. These are just as capable of generating organizational risks as the project themselves, so we will never be at the point of having completed risk management. Of course, we will never *start* it either—it's an ongoing process. At the program level, things are a little simpler, but risks will still run across multiple annual planning cycles.

This means we can't implement a simple project style planning and execution process for the portfolio or program; however, we still need to respect the basic rules of identify, then analyze, then manage. To meet this need, instead of a single major process and activity sequence with occasional adjustments (as we see in projects), the various elements of organizational risk management occur as shown in Figure 7.2.

This has a number of differences from a project risk management approach. The first is that it is a cyclical process—the contingency and impact assessment is not the end of the process as it often is on a project;

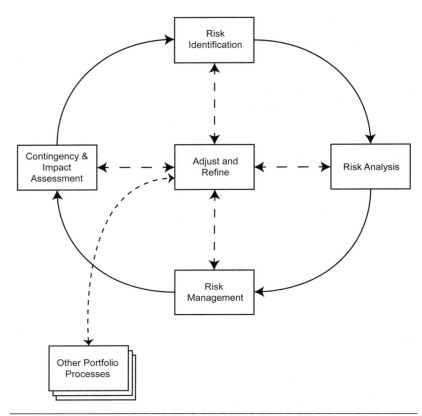

Figure 7.2   Sequencing organizational risk management activities

rather, it is a major source of additional risks. A triggered risk, by definition, changes the risk profile of the organization and that requires us to
review the profile to see whether the new *baseline* is driving new risks
(threats or opportunities), has eliminated previously identified risks, or
is changing elements of the remaining risks (likelihood of triggering,
impact, management approach, or exposure). More fundamentally, there
may now be issues with the balance between risk exposure and risk capacity, and if that isn't addressed, there may be substantial changes to the
risk management approach that are driven into the process. In the most
extreme examples, a triggered risk may change the constraints hierarchy,
which will force a comprehensive re-analysis of all of our risks and risk
management activities.

The second difference between this process and project level risk
management is that the individual steps will overlap a lot more than in a

project. At a project level, we tend to undertake a major risk identification exercise at the start of the project, which then leads to analysis, prioritization, and management. There will be additional risks identified during project execution (or at least there should be), but these will be a small percentage of the whole. At the organizational level, while there may be periods of higher and lower activity as major projects and programs kick off or wind down, there will be a regular stream of risks that are identified and passed on to the analysis processes—it's an ongoing exercise.

In the center of the process flow diagram, you will see we have an additional item for adjustment and refinement. This has two-way interactions with each of the four steps in the organizational risk management process flow and represents the fact that there is significant interconnectivity between risks at the portfolio and program levels. In the same way that contingency and impact assessment drives additional risk identification activities, so the actions taken for a risk at any of the steps in the process can have an impact elsewhere in the risk management of the portfolio. A reduction of risk exposure on one project may result in the elimination of another risk, or it may allow for additional management effort elsewhere, etc.

The same connection between risks happens at the project level, but the scale and complexity of the interconnectivity is significantly reduced there; a discussion in a status or risk review meeting is likely all that is needed to ensure that the full impact is understood. At the portfolio level, we need a much more robust process.

Related to the interconnectivity of risks, you will see that there is a connection between the risk management processes and the rest of the portfolio processes (which will include program management). This is an obvious connection—risk management doesn't exist in a vacuum, but it is still worth capturing. We connect to the adjust and refine element of the process because this is the best place for the determination of the impact that a change has on the risks or vice versa. The nature of the connections here are virtually limitless, risks are so diverse, and the portfolio is so wide ranging, that the potential for impact reaches virtually every department, process, and person within the organization. Clearly this requires considerable flexibility in the way that the interactions between organizational risk management and other organizational areas interact, but we still need a consistent process to manage the *gate* between risk management and the rest of the organization.

The final difference you will notice between the organizational process in Figure 7.2 and a traditional project level risk management approach is the absence of a process for planning risk management. That doesn't imply that we don't have to plan the way that risk management is carried out at the organizational level, in fact quite the opposite. It's true to say that there is more stability to organizational risk management—there's only one portfolio vs. numerous projects. As we have already discussed there is a cyclical nature to organizational risk management, rather than a specific *journey* through a series of process steps, and this does remove much of the need for planning of a specific execution of organizational risk management.

However, the main reason why we don't have a section on organizational risk management planning is that the concept needs to be built into everything we do leading up to an item becoming part of the portfolio. Consider how a project is approved for inclusion in the portfolio. It typically starts with an idea that is discussed among a few people, it gets enhanced and improved, and it is then presented to one or two leaders in the organization. There it may join a few other ideas that the management levels have developed themselves, and they are combined, refined, and consolidated down to a few select ideas. The best of those have business cases developed, which are again reviewed, refined, and finally submitted to the formal project selection process. At that point they may be modified or enhanced again, and if they are considered to be aligned with the organization's priorities and deliver sufficient return on the investment, they are approved and officially become part of the portfolio.

So at what point should we start thinking about risks for those initiatives? At a minimum, portfolio management takes responsibility for them once they are approved, and I would argue strongly for mandatory portfolio management engagement in the project selection process. I would go further and recommend the involvement of portfolio management during the idea generation and refinement (but that's a different book), so where exactly does risk management start? My answer to that is that we should start thinking about risks as soon as we start thinking about the idea—at the point where someone goes to see a colleague and says "I've got an idea that I want to run by you," they should be starting to think about risks, albeit without much formality. The tools, templates, and guidelines that the organization has for turning ideas into business cases should facilitate that thinking even though they aren't a part of risk management. As soon as the idea reaches the point where your organization considers portfolio management to have responsibility—during business casing, during review, or only upon approval—then that initiative should enter the first part of the organizational risk management process, the formal risk identification. There's no need for planning because the portfolio is already well under way and the initiative that is joining the portfolio is already prepared for risk identification because of the foundation work that has been completed.

One final comment before we look at the process elements in more detail. There will be differences between the specific elements of portfolio risk management and program risk management, and we will explore those in the chapters that focus specifically on those elements, but consider this approach as crossing levels between portfolio and program as needed. An organizational risk management process recognizes that the portfolio will drive risks and risk management into programs and vice versa. The portfolio and programs will also drive risks to the constituent projects, and we will explore that later in the book as well.

Now that we have seen the overall process, let's look at each of the elements in more detail.

8

Process Framework—Risk Identification

In project risk management, risk identification is a straightforward process. There are inputs to the process, but those inputs are fairly simple—the major project documents that have been developed at that early stage. At the organizational level, life is far more complex. As we saw in previous chapters, risk management can occur at different points for different projects, and risk identification will occur at different degrees of formality in those early stages before a proposal becomes an official project. This causes a lot more potential inputs to the process, and those inputs will not all be ready at the same time or to the same degree of detail. The inputs will also be updated frequently, especially during the business case development and review, and one of the outputs of risk identification may well be further updates to those inputs—that is, the risk identification process drives further change into the business case/project proposal.

Table 8.1 is a summary of the inputs, process elements, and outputs of the risk identification section of the organizational risk management process.

Inputs

The first input that we need to consider is the set of organizational goals and objectives—these form the foundation of everything that happens within the portfolio, risk management included. The portfolio only exists to deliver the business benefits represented by those goals and

Table 8.1   Risk identification process summary

Inputs	Process elements	Outputs
Organizational goals and objectives	Group and individual review	Updated organizational risk profile
Organizational risk environment	Clarification and validation	Updated portfolio/program risk lists
Organizational risk profile	Consolidation	Individual risk summaries
Organizational constraints hierarchy		Updated project proposals/business cases
Project proposals/business cases		Updated project plans (new risks)
Project plans		Updated actions log
Organizational project archive		
Portfolio/program risk lists		

objectives—success is not measured in terms of tangible project or program deliverables, but rather in the way that those deliverables contribute to the organization's overall success.

These goals and objectives will be relatively stable and will not directly drive much risk activity, but they are the crucial measure against which everything is judged. Remember that in the Introduction we defined a risk as something that has the potential to impact objectives—this is where we apply that measure. If an identified potential portfolio level risk will not impact these goals and objectives, then it is not a risk! We should also consider program risks against the same standards; the success of the program should still be measured against its goals and objectives.

The second input that we consider is the overall risk environment of the organization—the external risks that influence the organization. We looked at this in Chapter 1 and it helps portfolio and program managers understand the factors that may impact the organization and, by extension, the portfolio. This is another example of an input that doesn't directly drive actions. It's not even a material input in the form of a document, but it is a requirement for a level of understanding to facilitate effective risk identification—it's part of the larger knowledge base that will help to determine how real a risk is. Remember that portfolio management is a strategic function making critical decisions that impact the entire organization—those decisions require as good an understanding of the organization's situation as possible. You can't make the best decisions without the most complete information set possible.

As we consider the more direct inputs to the risk identification process, we first have the organizational risk profile. This is a great example of the type of item that benefits from the cyclical approach to organizational risk management. The organizational risk profile is a living document that will evolve over time; it is never *complete* but merely represents a snapshot of the risks that the organization faces at any given point in time. As such, it provides input to organizational risk processes and is updated by them. In the case of risk identification, we use the profile to help us ensure that all of the potential risk areas are considered and that the overall risk capacity and exposure is understood, as this too will drive risks.

This is complemented by the organizational constraints hierarchy that we considered in the previous chapter. This will be a significant input later in the organizational risk management cycle, but here it is an indicator of risk tolerance. The relative position of risk in the constraints hierarchy helps portfolio managers understand how aggressive risk management needs to be. We mustn't forget that the portfolio management function is accountable for the success of the entire portfolio, not just risk management! If risk is at the bottom of the hierarchy, then we should be spending less time on risk identification (and the other risk management processes) then if it is at the top of the hierarchy, as the implication is that there is a higher degree of risk tolerance. The portfolio level constraints are still applicable to programs that need to deliver against portfolio constraints. The constraints hierarchy of programs may be slightly different from the portfolio level hierarchy (if they are significantly different, then it likely indicates an issue) and may form additional program level inputs, however they cannot replace the portfolio level hierarchy.

One of the largest categories of risk inputs will be the individual project plans from the different initiatives that collectively make up the portfolio. This will form a lot of material, and the documents will not be studied in great depth, but they will be reviewed for the trends that can identify where risk exposure exists—over reliance on resources or vendors, assumptions that collectively increase risk exposure, etc. Where initiatives are not yet well enough advanced to have project plans, the project proposals will form inputs as long as this is considered part of portfolio management within the organization. I really do strongly urge organizations to view portfolio management as starting prior to the project selection process, as the function can add considerable value, but that is a topic for a different book. While care needs to be taken in putting too much weight on, or driving too many actions from, an initiative that is not yet approved, the risk identification process can both benefit from and drive tremendous value into the project proposal/business case process.

The next input, and one often ignored at all levels of the project execution organization, is the organizational project archive. This is the historic record of what has happened in the past and is often the greatest assistance available to portfolio and program managers. The quality of the archive will depend on the tools and processes in use within the organization, but it should minimally be capable of providing a record of the risks identified during previous portfolio execution cycles. Of course, those can't simply be transferred into the current risk identification exercise in a *cut and paste* format, but neither should the assumption be made that they have no relevance. Every situation is unique, but few have no areas of overlap.

The final input is simply the current risk list that has been generated by previous iterations of the risk identification process. This list will be appended and updated during the process.

Process Elements

Risk identification itself is a straightforward process that consists largely of producing a list of risk candidates and then determining which are real. In most cases, this will be an update to an existing list (as we identified in the inputs), although it is possible that organizations which operate on an annual portfolio cycle will produce a new list at the start of each portfolio cycle. That shouldn't be assumed however, as there will likely be some overlap from late running initiatives or from ongoing programs that are driving portfolio level risks. Programs themselves will go through the process of generating a new list when the program is initiated and will then enter the cyclical approach. The complete process for risk identification is shown in Figure 8.1.

The challenge in this element of risk management is not usually one of identifying risks, but rather ensuring that:

- identified risks are truly new rather than adjustments to existing risks
- risks belong at the organizational level rather than elsewhere

If risks are misidentified or misinterpreted, then significant effort can be wasted as risk identification will drive a lot of downstream activities. This really can be quite challenging because portfolio level risks are more complex than at the project level where the cause and effect tend to be

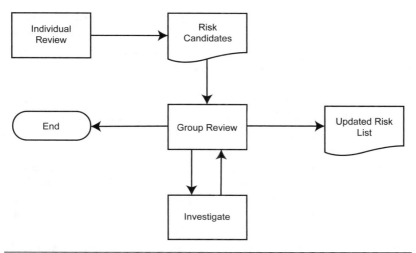

Figure 8.1 Organizational risk identification process

direct, which makes identification easier. We saw in Chapter 3 that project and program level risks can have impact at the portfolio level, but we need to make a distinction between organizational risks—the risks that we need to manage at the portfolio and program level—and project risks with organizational impact. We can't ignore the latter, but they are already being managed at the project level, so we need to treat them differently. We'll look at that in more detail later in the book. For now, an organizational risk is considered to be:

- additional to the project risks—it is a risk that has not been identified and managed by any of the individual initiatives that form the lowest level of the portfolio
- something that has the ability to impact the portfolio's and/or programs' goals and objectives (our test of whether it is really a risk)

Organizational level risks often occur when we roll up individual initiatives. We might find that a large number of projects are dependent on the same vendor, or that there is a bottleneck created in a limited resource by the current scheduling of work. There will also be additional risks that are not directly attributable to an individual project or combination of projects.

Suppose for example that a number of projects within the portfolio have revenue targets associated with them. If those targets are conservative, then there may not be much concern about meeting the overall portfolio revenue target. However, if all of those project level revenue targets are aggressive, then there may well be a significant risk of the portfolio missing its revenue objectives because success will require every revenue generating project to be executed without any issues—an unrealistic expectation.

It is important to have the right people involved in the risk identification activities, and while it can be tempting to engage project managers to bring their expertise to the work, they may not have the right perspective to contribute. They likely don't have visibility into the full breadth of the portfolio and programs, and they may find it difficult to retain an objective viewpoint. They will always be tempted to look at things from the perspective of their own projects. Project managers can certainly add value to the validation of risks, and they are vital to understanding the higher level impact of project level risks, but care should be taken if giving them a role in the initial risk identification. Ideally, identification should be confined to the portfolio and program management group, along with any PMO resources who are focused at the portfolio and program levels, although that's not always feasible as portfolio and program level resourcing is generally low.

Each individual involved in the process, regardless of his or her background, skills, and experience, needs to answer this question: "*Based on my understanding of the portfolio, my study of the factors that affect the portfolio and programs, and my skills and knowledge with risk management, what risks exist that have not yet been captured in the risk list?*" Initially this should be conducted as an individual exercise to allow for a more comprehensive review from multiple perspectives. It can be tempting to allocate different potential risk sources to different people for the sake of efficiency, but risk identification is not an exercise in efficiency. The goal has to be to accurately identify, as close as possible, 100% of the risks, and that is more likely to occur with more people reviewing the inputs for risk candidates.

Once the individual analysis has been completed, they are reviewed in a group setting by all of the participants in the process. This review will provide an opportunity to question whether items are in fact risks, whether they are new or variations of existing risks, and whether they are at the appropriate level. These discussions have the potential to become animated, and while this is a good thing in terms of coming up with the best solution, it can also lead to some interpersonal challenges, with the potential for bruised egos and hurt feelings. An independent facilitator is recommended to help keep the meeting objective and focused—a senior PMO resource is a good example. The meeting also needs to be conducted as a group of equals; submissions that carry more weight because of who presents them rather than because of their merit will not result in effective risk identification.

These discussions may take multiple sessions to complete, and this investment of time and effort is worth it. We shouldn't limit risk identification because of an arbitrary time limit. That said, if discussions over a particular risk candidate cannot reach consensus in a reasonable amount of time, then that item should be documented and put to one side. Once the discussions have been completed, those risks can be assigned to specific participants or groups of participants for further analysis so that a final decision can be reached. The need for this additional work is almost a certainty given the far-reaching elements and impact the portfolio and programs have. If there are no such outstanding items after a group review, then it may be an indication that decisions have been made based on dangerous assumptions—putting speed ahead of thoroughness.

Every such research item should be assigned to a specific individual owner (even if a group is going to be involved) and given a date to provide an update on progress. The new risks that were confirmed can continue with the risk analysis processes, and the ones that remain uncertain can either be resubmitted as updates during the next risk identification discussion (remembering that this is an ongoing process at the organizational level) or reviewed in a separate meeting convened for that purpose.

Outputs

The outputs to the risk identification process are straightforward. The purpose of the process is to identify new risks, so the key output will be individual risk summaries—a basic overview of the risk, the potential impact, the reasons for concern, etc., along with the initial owner of the risk. This will then feed the analysis process elements that we will look at next. While many of these will be new risks, the discussions may also cause us to re-assess existing risks. In this case, existing risk summaries will be updated and resubmitted to the risk analysis process for re-evaluation.

In addition to the risk summary for each individual initiative, there will be a number of updates to the documents that formed the inputs to the process. The organizational risk profile will need to be updated to ensure that it reflects the current risk exposure, although there will be a need for further updates once the analysis of the new risks has been completed and there is a better understanding of the extent of the exposure. The portfolio risk list will need to be updated with the additional identified risks, but there may also be updates to the project plans if some of the risks identified through this process were determined to be real risks, but at the project level. These risks, which come out of the group review process in Figure 8.1, should be routed to the relevant projects for further analysis along with their individual risk summaries.

As we indicated above, there will also be updates to project proposals or business cases that are being prepared for submission to the project approval process if they are included in portfolio management. This will help ensure that the submissions are as complete and accurate as possible and supports better more informed decision making in the approval process. Finally, in the list of outputs, the actions log should be updated with all of the actions, owners, and due dates that have been created by this process. This will include both the work sent to the organizational risk analysis process and the work sent out to individual projects where confirmation of the integration into the relevant risk management process is required to complete the action. There may also be additional one-off action items that have come out of the risk identification exercises, and these too need to be captured or they risk being lost (without being noted as a risk).

9

Process Framework—Risk Analysis

Risk analysis is vital for successful risk management as it is the point where we plan how to deal with the risk. Analysis is to risk management as planning is to project management, if you will. Critical determinations must be made at this point:

- How likely is the risk to trigger?
- How significant is the impact if the risk triggers?
- How easy is the risk to manage (capability and capacity)?
- What is the right management approach?

If the wrong decisions and conclusions are made here, then not only can it result in the risk exposure being increased unnecessarily, it can also result in resources being committed to work that is adding no value to the organization.

Inputs

We concluded the risk identification process by considering the outputs to that element, and as you would expect, a number of those outputs become inputs to the analysis process. Table 9.1 summarizes the inputs, process elements, and outputs for risk analysis. The individual risk summaries are the lowest level unit that *carries* an individual risk from identification to analysis, and these form the major inputs to the process. We also need to consider the risk list that was updated upon completion of

Table 9.1   Risk analysis process summary

Inputs	Process elements	Outputs
Individual risk summaries	Qualitative analysis	Updated portfolio/program risk management plans
Portfolio/program risk lists	Quantitative analysis	Updated portfolio/program risk lists
Portfolio/program risk management plans	Prioritization	Updated organizational risk profile
Organizational risk environment	Determination of management strategy	Updated individual risk summaries
Organizational risk profile		
Portfolio constraints hierarchy		
Portfolio resource allocations		
Project risk management plans		
Project proposals/business cases		
Organizational project archive		

the identification processes, as this provides us with the summary of all of the portfolio level risks.

Risk analysis also shares a number of inputs with the risk identification process—the organizational constraints hierarchy is an important input, as are the organizational risk profile and risk environment. These will all influence our analysis—we cannot consider the impact of the risks in isolation, but do so in the context of the overall risk. These inputs all contribute to our understanding of that environment.

In addition, we consider the project level risk management plans here as a source of potential assistance in analyzing the risks. There may well be similar risks that have already been considered lower down in the portfolio, and that work can make the organizational risk analysis easier. We also consider the project proposals/business cases (subject to the caveats described above about the scope of portfolio management). While these won't provide as much detail on the risks related to them, they can help us understand the nature of the risk.

The organizational project archive can be just as helpful here as it was in the risk identification process. Few completely new risks will be encountered, and the more that we can learn from history, the more reliable our analysis will be. This archive can help us understand how similar risks were assessed previously and can provide a valuable framework for our analysis. However, we still need to interpret and process the data from lower-level plans and the archive. We have to validate that the historic scenarios can be applied, and we need to adjust for the inevitable differences. We also introduce two new inputs to this process that we have not previously seen. The portfolio and program risk management plans are important inputs, as they will help us in determining the best course of action to take with the risks once they have been analyzed. The risk management plans are not the same as the risk lists—they are much more detailed and cover only a subset of the risk list that is being actively managed. If this is the first time that risk analysis is being performed, on a newly initiated program for example, then there will not be an existing risk management plan; but in all further iterations of the process, this will be an input. We'll look in more depth at the risk management plans later in the risk analysis section.

The final input is the portfolio resource allocations. This should include the allocation of resources across the entire portfolio, including within programs and projects. This input provides a valuable insight into the skill sets and time available within the entire portfolio resource pool, which in turn will help determine how best to allocate ownership and determine management approach.

Process Elements

Risk analysis is the most complex of all of the organizational risk management processes. It has four main elements to it:

1. *Qualitative risk analysis.* The subjective review of each of the identified risks in order to understand the likelihood of occurrence, the impact if they do occur, and the cost to manage.
2. *Quantitative risk analysis.* The objective review of the risks that adds value and clarity to the qualitative analysis.
3. *Prioritization.* The sequencing of risks from *worst* to *least bad* (or *best* to *least good* when considering positive risks) in order to help determine which risks will be actively managed and which will be passively managed.
4. *Determination of management approach.* Even though this is the analysis process and not the management one, this is where we determine the most appropriate approach to managing the risk.

In addition, there needs to be a review and approval gateway that helps ensure that the analysis is accurate and complete. The downstream implications here are significant, and there needs to be safeguards in place to confirm that the management activities executed are appropriate and sufficient for each risk. The overall risk analysis process is shown in Figure 9.1.

Risk analysis is conducted for each individual risk summary that comes out of the risk identification process, but the analysis also considers the new risks as a group, as well as considering them against the universe of existing risks. In theory, the identification processes should have ensured that any risk duplication (examples of different symptoms of the same underlying risk) were eliminated before they left that process, but in reality, that may not always be possible. We may need to conduct some of the risk analysis work before we are able to conclude that multiple apparent risks actually have the same source. There is also always the potential for errors to occur, and having checks and balances built into our processes, such as looking for duplication in both identification and analysis, helps to minimize the impact of those inevitable errors.

We also need to recognize that risk analysis is not a *one-and-done* event. Risk summaries enter the analysis process once they are created in risk identification, but they may reenter the process a number of times during their life as they change and evolve, as the portfolio changes, and as the organizational risk environment changes.

Before we look at each of the process elements, let's ensure that we understand exactly what the analysis is designed to determine. At a minimum, risk analysis is concerned with determining the following:

- The likelihood of the risk occurring
- The impact that the risk will have on the portfolio and/or programs if it does occur
- The effect that management will have on the risk exposure (likelihood to occur and/or impact)
- The effect that management will have on the organization—will we be moving risk exposure elsewhere (see discussion below on risk elimination) or compromising another constraint to protect risk
- The cost of management
- The ability of the organization to manage the risk

Only when each of these is understood can we make an appropriate decision about if and how to actively manage the risk. Analysis inevitably focuses on trying to distill the risk down to values—management costs, impact costs, etc. However, we can't lose sight of that last bullet point— the ability (or inability) of the organization to manage the risk is a vital

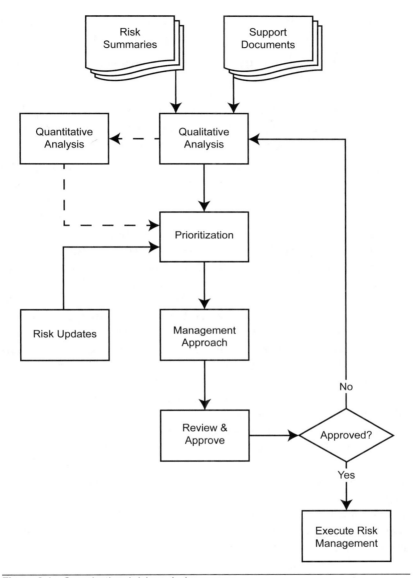

Figure 9.1 Organizational risk analysis process

consideration. If we can't reduce the likelihood of the risk occurring or the impact that it will have if it triggers, then we simply have to accept the risk and its potential impact and move on, or we have to modify the portfolio in order to eliminate the risk.

The risk analysis process begins with qualitative and quantitative analysis. Books have been written on just this subject, but put simply:

- Qualitative analysis applies expert judgment and interpretation of the data that is available in order to determine a non-exact, subjective interpretation of the risk.
- Quantitative analysis builds upon qualitative analysis to apply objective criteria to the risk in an attempt to determine a more precise (but not necessarily more accurate) interpretation of the risk.

Qualitative analysis should always occur first as it provides the principles upon which quantitative analysis can be applied. An example may help to clarify.

Suppose we have determined that the portfolio has three individuals who are critical to success—without them the portfolio will fail. That has led us to identify a risk that those people may leave the organization, with obvious dire consequences. Our qualitative analysis involves us using our knowledge of those people, the amount of time that they have been with the company, the role that they have, and their personalities and attitudes. As a result of that, we have determined that we have a 10% chance of one or more of them leaving before their work on the portfolio is complete.

Our quantitative analysis involves us looking at statistics of the average tenure that people in similar roles have had within the organization over the past 20 years, the relationship between number of days absent and likelihood to leave, their last performance review and the likelihood of people with similar performance results to remain or leave, and a myriad of other potential data points. We determine that based on the number crunching, there is a 12.76% chance of them leaving. We should never rely solely on objective, quantitative criteria, but we may wish to adjust our 10% estimate based on the higher quantitative number. However, also consider the earlier statement—quantitative analysis gives us more precise estimates, not necessarily more accurate ones. We may decide that our interpretation of the risk, the qualitative analysis, is more appropriate in this case than a statistical analysis that ignores the uniqueness of the individuals.

Note that qualitative/subjective analysis does not mean that we don't have a numerical estimate. We need to apply numerical values to our risks—management costs and impacts for example, because these are needed throughout the process. We ultimately need to be able to distill our risk exposure down to estimated financial and time values. As we saw when we built our organizational risk profile, every element is dependent on numbers, but these don't have to come from a quantitative analysis.

They can be generated as estimates from our qualitative analysis, and we can use the management reserve to reflect the level of confidence that we have in those estimates (the greater the confidence, the lower the variation from the estimate should be and the less management reserve is required).

We always have to consider qualitative and quantitative analysis in partnership because they are so closely tied together. However, as you can see in Figure 9.1, quantitative analysis is tied to the rest of the process flow with dotted lines. This is because it cannot always be undertaken, and where it is carried out, it may not be possible to apply it to every aspect of the analysis. This is both acceptable and inevitable. While it is always nice to be able to quantify the risks that we are facing, it is simply not possible in many situations—how do we quantify the impact of the takeover of one of our suppliers for example? Quantitative analysis should always be considered as a *value add* to the process, not a mandatory step.

The analysis process elements will always involve a trade-off between effort and reward. The more detailed our analysis, the more accurate our numbers should be, but the higher the cost of conducting the analysis. We can't simply apply an arbitrary limit to the amount of effort that is spent on risk analysis because every risk is different, so we have to start our qualitative analysis with a preliminary interpretation of the likely exposure that we face from the risk. This makes the resourcing of the qualitative analysis critical. The people involved need to be able to apply their skills and experience to the process and ensure that the best possible analysis is conducted. By definition it's a subjective process and relies on the ability to interpret the available information, reach conclusions, and apply judgment, something that can't be done unless the person has considerable risk management experience.

This experience needs to be complemented by an understanding of the unique considerations of the portfolio or program for which the risks are being considered. This is one of the reasons why inputs like the organizational risk environment, the organizational risk profile, and the organizational constraints hierarchy are important. We need to look at these as more than just specific inputs to specific analyses. It's important to be consistent in the execution of risk analysis, and a discussion of the portfolio/program drivers and the organizational situation between the people involved in that analysis can help to ensure that there is consistency in:

- Understanding the elements of the portfolio or program that are exposed to the most risk—which goals and objectives are the most vulnerable, for example.

- Understanding the areas where the organization is prepared to be more risk tolerant and where it is more risk averse.
- Determining the preferred management approaches (although analysis is undertaken prior to the management approach being determined, an understanding of available and/or preferred approaches to managing the risks helps us to understand the potential impact).

The analysis may well start with a high-level review of the risks that need to be analyzed. While a decision to rely on a simple cursory examination followed by an assumption that this is a *standard* risk is dangerous, it may be necessary if the management team is going to be able to deal with the volume of risks that enter the analysis process. The costs of a detailed analysis of every risk may simply be too high and may not always add any further value—the preliminary conclusions will be accurate. The skill comes in determining which risks are not standard and do require a deeper level of analysis.

Where more detailed analysis is necessary, analysis will begin with the inputs to the process that we identified above but may need to evolve into additional work that is unique to the particular risk that is being analyzed. We should always consider the organizational constraints hierarchy to help us understand the trade-off between money and effort that can be spent on the analysis. We must also ensure that we don't reach the point of diminishing returns—where we will spend more money and time in understanding the nature of the risk than we will be able to save the organization as a result of that deeper understanding.

Once analysis of the risk is complete, we need to prioritize it against the other risks that impact the portfolio and programs. This is another process that many will be familiar with from project management but where the process mechanics are different at the organizational level. Prioritization has to recognize that the list of risks is fluid—new risks are added on a regular basis, and risks are retired from the list as they are eliminated by progress on the portfolio and programs, or the risk triggers. Additionally, the risks that are on the list will themselves be changing and evolving, and the risk owner will be updating them as part of the management process that we will look at next. Theoretically, the constraints hierarchy and organizational risk profile will also be changing and impacting the prioritization, but realistically, a significant shift in one of those areas will trigger a re-analysis that in turn will drive the changes to priorities.

We also need to have a reasonable expectation for what prioritization is designed to achieve at the organizational level. A portfolio is likely to have a relatively large number of risks, and while we need to prioritize those risks in order to ensure that we are focused on the correct ones, we also need to recognize that the sequencing of 100 risks in exact order from most severe to least severe is a pointless exercise. A debate about whether a risk is 53rd or 54th on the list is not going to lead to better risk management (even ignoring the fact that the risk list will have changed again within a short period of time)!

At the organizational level, we should therefore view prioritization as an exercise in categorizing risks by the degree of attention that they require. We'll get into the management activities themselves when we get to the risk management processes, but the categories that we use (critical, severe, serious, intermediate, and minor are one possible categorization) should be clearly defined for the organization. Then the implications of categorizing a risk into a specific category are clearly understood and consistency exists across the portfolio and programs (and projects).

To get to this point in the risk analysis we have used a combination of judgment and measurement, and we need to do the same in the prioritization exercise. Many organizations will have purely quantitative measures to prioritize risks—a mathematical combination of the impact and the likelihood of triggering. This creates a direct connection between the contingency reserve and the prioritization, where the more substantial the contingency reserve, the higher priority the risk. While this has some value, it eliminates the value that experience and judgment brings to the process. Consider the following example to illustrate the point:

- Risk A has a 50% chance to trigger and, if it occurs, it will cause a three-month delay and a financial cost of $100,000.
- Risk B has an 80% chance to trigger and has the potential to cause a six-month delay and a $250,000 cost impact.

Which risk is the more severe? It seems obvious—Risk B is higher in every measure so must be a more severe risk, right? Certainly it would be if we only used quantitative measures in our prioritization, but let's consider some other factors. What about if Risk A has been actively managed for several months with no material improvement in the risk parameters, but Risk B was only identified a week ago and is already showing signs of responding to our management approach and the exposure looks like it may soon reduce. That doesn't automatically make Risk A more severe, but it's another element that we should be considering, and that requires the inclusion of subjective measures in our prioritization activities. Similarly, if Risk A was only a 20% chance to trigger with $40,000 in impact a week ago, then clearly it is becoming more severe, and this needs to be considered when comparing the two risks.

An effective prioritization process should consider all of the following, as well as the numerous unique characteristics of each individual risk:

- Exposure—effort and cost
- Likelihood to trigger—as we saw with the risk profile, risks with low chances to trigger and high potential exposure can sometimes give skewed indications of the real risk if we only look at objective measures
- Ability to manage—both whether it can be managed and whether resources are available
- Cost to manage
- Trend—whether the elements of the risk are improving or worsening
- Response—how the risk is responding to management—this is different from trend because it focuses specifically on the success or otherwise of management, whereas trend may be the result of a number of different factors
- Exposure area—the elements of the portfolio and/or programs that will be impacted by the risk if it triggers and the areas that are incurring the management cost
- Timing—whether the risk is in imminent danger of triggering or the danger is many months away
- Owner's judgment—the view of the risk owner and/or the person conducting the analysis as to the severity of the risk

The results of this analysis will be a categorization of the risk into one of the predefined categories (severe, intermediate, etc.), and the individual risk summary should be updated to include a summary of why that determination was reached—especially if the conclusion is different from what is indicated by a purely quantitative calculation of severity.

Once this has been completed, we can move on to the process of determining the appropriate management approach. We briefly looked at the four different approaches for managing negative risk in Chapter 4 (mitigation, elimination, transference, and acceptance), and at a fundamental level, we will be applying one or more of those strategies to each of the risks that we need to manage. The details will be a lot more complex than that, but those four fundamental approaches are a good place to start.

As a result of the prioritization, we have already identified many of our candidates for acceptance of the risk. Acceptance is a passive risk management approach—we will not be taking any active measures to try to control the risk at this point. The most obvious candidates for acceptance are the lower priority risks—those given the lowest of ratings on whichever severity scale the organization is using. These will generally be the risks where the exposure and/or the chance to trigger are relatively low and where the benefit that could be delivered to the organization from a more active management strategy would simply not be worth the effort.

It's important to recognize that acceptance is still a valid risk management strategy—we are consciously determining that at this point the risk does not warrant a different approach. The risk will still be assigned an owner and will still be subject to review on a regular basis to ensure that the acceptance strategy remains appropriate.

Once these *easy* risks have been handled, there will likely still be a relatively high number of risks remaining, and this is where decisions about the management approach become more difficult. Our analysis will have provided an estimate of the costs to manage each of the risks we are facing, and we will have an understanding of the capacity for managing risk within the portfolio from the organizational risk profile. Unless we are extremely fortunate, these two numbers will not be aligned. Either we will have excessive management capacity, or, more commonly, we will find that the costs to manage the risks exceed the available resources for those management activities—we simply don't have enough money and/ or people available to manage all of the risks effectively. Additionally we have to determine how those resources should be distributed—how much effort and money is focused at the portfolio level and how much is assigned to programs, projects, and elsewhere in the organization. Remember that the organizational risk profile represents risk management capacity for the entire organization, not just for portfolio execution.

This is where the analysis and prioritization need to come together in order to ensure that we are investing our time and money where it will have the most significant benefit to the organization. Let's start with our most severe risks and determine whether we absolutely cannot live with any of them, even if they can be managed. If some risks have the potential to so severely impact the portfolio, or a program within the portfolio, that they must be eliminated completely, then those have to be dealt with first. In many ways these are low-cost risks—we aren't going to be spending time and money on them for an extended period. However, there is little chance of a risk being completely eliminated without making a fundamental change to the portfolio and/or program—we have to eliminate the source of the risk, and that means removing that element of the portfolio. The impact of this change will be part of the analysis that was conducted. It will have established whether it is *just* a matter of changing the approach to one of the project or program elements (building in house rather than rely on a risky supplier for example) or whether it will require changes to project and program deliverables and potentially an inability to achieve the portfolio goals and objectives.

These are not easy decisions to make, but they must be made decisively in order to maximize the chances to recover the overall portfolio and minimize the time, effort, and money spent on a risk that is ultimately eliminated anyway. When it comes to elimination of a risk, there should also be an understanding that it will have an effect—elimination generally involves a fundamental change—removal of a portfolio element and/or a significant change in approach. Either of these will drive additional risks into the portfolio through a combination, making it harder for the organization to achieve its goals (by moving away from the preferred portfolio makeup) and adding new activities that will themselves drive risks associated with that work (our building in-house vs. outsourcing example).

Elimination as a management strategy needs to be viewed from an organization-wide standpoint. We may be eliminating a specific risk, but across the entire portfolio we are likely simply moving the exposure to areas that are deemed to be more acceptable, and in some cases we may actually be increasing the total exposure as a result of these actions. There is nothing wrong with this approach, and the analysis should have identified the likely impact of an elimination strategy if it was considered appropriate, but we still have to consciously acknowledge that this shift is occurring. A decision to eliminate a risk may also drive new work into the risk identification or risk analysis processes to formalize that impact.

The next category of risks to consider are the ones that our analysis shows will not respond well to any attempts to manage. These may be at any level in the prioritization, but because of the unique nature of the risk, the application of money and/or resources to managing them will have no material impact on the exposure or chance to trigger. As an example, perhaps we determine that using the risky supplier discussed here is not so severe a risk that we need to eliminate it, or perhaps we have no other options and have to use that supplier. However, we know from the attempts that we have made to manage the relationship over the previous several initiatives that there is little we can do to influence the supplier to better meet our needs, so we apply minimal effort to managing the risk but simply use our acceptance strategy. It may not be a decision that we like making, but sometimes we have to accept it and move on.

The decision to accept risks in this category is different from the decision to accept the lower priority risks that we discussed earlier. In that case, we consciously decided to accept them because the risk severity allowed for it. In this scenario, we are being forced to use the acceptance approach because nothing else will work. These risks still get owners, and they are still reviewed regularly to see whether the situation has changed, but otherwise they don't have any resources assigned to them.

Once these different categories of risks have been handled, we will be left with a group of risks that spans a large percentage of the overall prioritization and the two major strategies remaining—transference and mitigation. Let's deal with transference first as that's the simpler management strategy to address. By definition, transference involves moving responsibility for the risk outside of the portfolio. This is likely to result in either new risks being created from the inclusion of that additional party, increased exposure for existing risks (greater reliance on a vendor for example), or both. The analysis should have identified this likely additional exposure, but it still needs to be considered in the overall decision about whether transference is an appropriate strategy.

One further consideration when we use transference—it is usually only a financial strategy. When we transfer responsibility for a risk to another party, it usually involves the payment of a fee—a contract with a supplier or an insurance policy. There is only a small amount of effort necessary. We may need to find the supplier and negotiate a contract, but generally speaking we are paying for the ability to make someone else responsible for the risk.

Similarly, when we consider the implications should a transferred risk trigger, we will still likely be faced with the effort exposure—transference only protects us from financial impact. For example, suppose that we negotiate a contract with a contractor to refit a floor of our building to allow us to move additional staff into the space. We could have used our own maintenance staff, but they aren't as experienced. We need to make sure that the work is done before December 31 because the lease on the building that those staff are currently using is expiring. As part of the contract, we insist on a completion date of November 30 (to provide a safety margin), with a penalty of $1,000 per day that they are late. If the work isn't completed until mid-January, then the contractor is faced with a penalty of around $45,000, but we still have to deal with the problem of some of our staff not having an office for a two-week period—that risk can't be transferred.

Transference is still a valid strategy. It is likely to be a better way to control the financial risk exposure than our other approaches (except for elimination), and it is also the most predictable form of risk management—we pay a fixed or semi-fixed amount in order to remove a fixed amount of financial exposure from the project.

The final approach is the one that will form the vast majority of the risk management activities that we perform—mitigation. This is the type of risk management that most people will be familiar with from project management—the use of resources (financial or effort) in an attempt to either reduce the chances of a risk occurring, the impact that the risk will have on the portfolio if it does trigger, or both. While we group these activities together under mitigation, the specifics will be tremendously varied from risk to risk and will be subject to a large number of variables. We'll look at how those variables affect the specifics of risk control when we get to the risk management process elements, for now we simply need to ensure that mitigation is applied to the right risks. There is a danger that mitigation becomes the *default* risk management approach because many risks can't realistically be eliminated. We don't want to just accept it (which still carries the stigma of *ignoring the risk*), and there isn't an opportunity to transfer. In projects, that's not too dangerous an assumption, but at the organizational level, this can be a dangerous approach.

Portfolio and program risks have the potential for significant impact—by definition a portfolio risk has the potential to affect the organization's ability to achieve its goals and objectives. Given that mitigation is an internal management approach, we keep control of the risk within the organization. If a significant percentage of the portfolio's risk exposure is retained within the organization itself, then that itself drives risks—it's no different than a heavy reliance on a single vendor in that it focuses the responsibility for success (or failure) in one place. When that focus is on our own organization, there is more control over the risk than if it is with a vendor. Yet we do need to recognize that we are creating a situation where if things don't go smoothly we potentially have to divert an increasing number of internal resources to risk management, or accept an increasing amount of exposure. That's not necessarily something to avoid, and in most organizations, mitigation will be the dominant risk management approach, but we do need to ensure that it is a risk that is understood and consciously accepted.

During the risk analysis process elements, we don't have to determine the detailed mitigation approach, but we do need to ensure that any risk flagged for mitigation is likely to respond in a favorable way to that management—and again our analysis should identify that. Mitigation for the sake of *feeling like we're doing something useful*, without any realistic expectation of delivering a tangible benefit, is simply wasting resources that could be better used elsewhere.

The final consideration here is the use of multiple types of risk management approaches—generally two of acceptance, transference, and mitigation (and occasionally all three). Transference in particular should be used with an additional approach by default as it is typically only of assistance in controlling financial risk and perhaps the likelihood of triggering, so we need to consider a strategy for the effort exposure.

Once we have conducted the qualitative and quantitative analysis, have prioritized our risks, and determined the appropriate management strategies, then we are almost ready to hand off the risks to the risk management process area. However, there is one final step that needs to be conducted. Risk analysis drives all of the actions that we will be taking to manage each of these individual risks, and any mistakes that are made at this point can have considerable implications downstream. Not only will we be investing time and effort in areas that are not generating a return on that expenditure, we are leaving the portfolio exposed to higher levels of risk because we are not managing them effectively.

For these reasons, it's important to have a formal review and approval process as a gatekeeping step before a risk leaves analysis and enters management. This doesn't need to be a committee meeting to discuss each risk analysis, in fact that's not desirable—it will be inefficient because it forces risks to *wait* for the next meeting before being approved. Additionally, in most cases the analysis doesn't require discussion; it simply requires an independent review and confirmation. The most effective approach will simply be a peer review by a resource who has not been involved in the analysis of the risk to this point. They are not trying to second guess the work that has been completed but will be focused on the following questions:

- Is the qualitative and quantitative analysis consistent with the risk that has been identified?
- Is the prioritization appropriate for the exposure that has been identified?
- Is the management approach/combination of approaches the most effective for the individual risk?

If the answer to all of those questions is yes, then the risk can move on to the management process. If not, then it needs to be re-analyzed, potentially all the way back to the initial analysis, although if the only question is around the management approach, then it can move swiftly to that point for reconsideration. Frequently all that is needed is a conversation between the person who analyzed the risk and the person who reviewed it, and perhaps some clarification of the documentation.

Before we look at the outputs of risk analysis, let's spend a few moments considering resources. Risk analysis can be a resource intensive activity. It can be tempting to assign one member of the portfolio management team to conduct all risk analysis, and this does have some appeal as long as that person has sufficient understanding of the portfolio and each of the programs within the portfolio to be able to conduct an effective analysis. It helps to ensure that consistent criteria are being used across risks, which makes it easier to compare risks when determining the priority. However, there will be times when there is simply too much risk analysis to be conducted for one person to be able to complete the work.

For example, upon completion of the annual planning cycle, a whole wave of new and modified risks will be driven into the process. If only one resource is assigned, then there will be bottlenecks, which will result in extended periods where risks are not being properly managed. This is in itself an additional organizational level risk, and there needs to be a willingness to add additional resources when needed. At the least there needs to be a second person who is always available for the review and approval process.

The problem comes from the fact that portfolio and program levels of the organization rarely have multiple resources assigned, and so there simply won't be many people available to assign to the work. That's not necessarily a problem; portfolio and program management are not resource intensive functions for much of the time because they act through projects, but there should be a plan to add short-term resources in scenarios such as this. These available additional resources should be selected based on risk management skills and experience rather than job title, but commonly they may be found within a PMO or project control office or among the more senior project managers within the organization. Where specialists are required to assist with elements of the analysis, that expertise should be pulled in from wherever it is needed, regardless of whether the individual has any other involvement with project related work.

Outputs

The completion of the risk analysis process generates a number of outputs. Because of the nature of the process, the outputs are updates to existing documents rather than completely new artifacts. First, we have updated portfolio and program risk lists, although these should be fairly stable if we have done a good job at the risk identification stage. If there are significant updates to your risk lists during the analysis stage, then it may be an indication that there is a problem with the risk identification work (process problem or execution problem). We would expect more substantial changes to the portfolio and program risk management plans to reflect the results of the analysis, the prioritization, and the high-level management approach.

Second, we will also have an updated organizational risk profile that reflects the overall portfolio risk picture and total exposure from the programs and the portfolio itself. This may include the identification of shortfalls in risk management capacity, either on the management side or in the reserves. Finally, the individual risk summaries will also have been updated with the detailed analysis that has been completed and the management approach that has been approved, and these will represent the largest amount of the work that comes out of the process.

10

Process Framework—Risk Management

Risk management represents the point where we make a tangible difference to the organization's risk exposure through the application of money and/or effort. The work that has been completed in risk identification and analysis will have been wasted if it is not applied effectively at this point. At the same time, there needs to be realistic expectations about what can be achieved in risk management. There will be a number of risks that do trigger, and there will be real impact on the programs and portfolio. This doesn't represent risk management failure, which will always be a balancing act of risk vs. return. The focus should always remain on the performance of the portfolio as a whole and the ultimate success of entire programs, not the individual risks within them.

The risk management process will control the risk from the completion of analysis to the point where one of the following situations is reached:

- Risk management has been successful, and no further work is necessary
- It is determined that additional risk management will have no tangible effect on the risk
- The risk has triggered

The risk management process is shown in Figure 10.1.

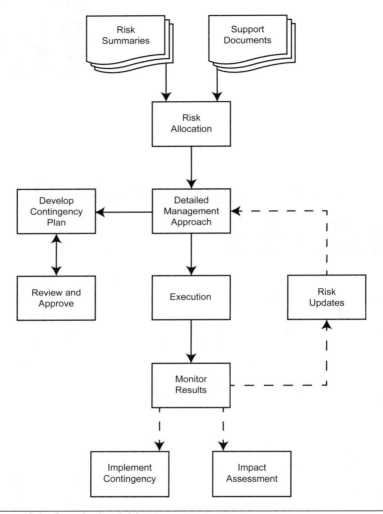

Figure 10.1 Organizational risk management process

Inputs

The inputs to risk management are straightforward. The two key sets of documents come straight from the outputs of risk analysis—the risk management plans at the portfolio and program levels and the individual risk summaries. The individual summary for each risk will provide the full details of what the risk is, the analysis that has been conducted on it, and the management approach that has been approved. The summary risk management plans will provide the overall picture of active risk management that needs to be executed for a program or the overall portfolio. The portfolio and program level risk lists are also an input to the process to provide the total view of all risks that exist on the portfolio, but these are less immediately important as passively managed risks will not be included in our active risk management activities.

The organizational constraints hierarchy is again considered here, as it is the key to ensuring the right decisions are made regarding the compromises. Finally, the portfolio resource allocations represent the availability of individuals within the overall pool of project execution resources. These are vital to ensuring that the available resources are utilized as effectively as possible in managing the risks without prolonged over allocation, and will also help identify where opportunities may exist to expand or adjust risk management activities and improve the overall effectiveness.

Table 10.1 summarizes the inputs, process elements, and outputs of the risk management process.

Process Elements

Risk management can be a resource intensive exercise, and like any other work element it's important to have the right people assigned to the right tasks. When risk summaries are received from the analysis process, they need to be reviewed and assigned a risk owner. The choice of owner will be based on a combination of the analysis and the availability of resources. Between the two sets of inputs, there will be information on the following:

- The elements of the portfolio that the risk impacts, and by extension the people within the portfolio who are currently best

Table 10.1   Risk management process summary

Inputs	Process elements	Outputs
Portfolio/program risk management plans	Risk allocation	Updated portfolio/program risk management plans
Individual risk summaries	Determination of detailed risk management approach	Updated individual risk summaries
Portfolio/program risk lists	Develop contingency plan	
Portfolio constraints hierarchy	Risk management execution	
Portfolio resource allocations	Response monitoring	
	Approach adjustment	

positioned to monitor and manage the risk. This may be related to one or more specific programs, to a specific business function (procurement, training, HR, etc.).

- The similarity to existing risks that are already being managed. This will identify any opportunities to leverage work that is already being done and minimize the net new work that needs to be conducted. We are still focused at an organizational level at this point—we aren't looking to consolidate management with project level risk management activities yet.
- The resources who have capacity to be assigned to additional work. This also needs to consider the period of time that they will be available for and the skills they possess—just because they are available doesn't mean that they are the right people.

It is unlikely that there will be a nice easy fit where the person with the ideal combination of skills and experience is working in the area that is impacted by a new risk, and has the bandwidth to take on responsibility for the new work. There will need to be an element of compromise in the decisions that are made, and there may also need to be some reallocation of resources in scenarios where it is considered imperative to have the best possible owner assigned to the risk—a situation that usually occurs with the most severe risk exposures.

This is where the organizational constraints hierarchy factors in. That can help us determine how resource allocation compromises should be made by identifying where risk fits relative to the other constraints. For example, suppose we have a risk that impacts an area of the portfolio where all resources are fully utilized and it will continue that way for some time. Giving responsibility for the risk to one of the existing resources is not a reasonable option—conscious over allocation of resources is not sustainable and will inevitably result in corners being cut sooner or later. There are several options at this point:

1. Pull people off of project schedule tasks in order to assign them to risk management activities
2. Give the risk a nominal owner on the relevant team but without any expectation of active management occurring
3. Give the risk to a resource elsewhere in the portfolio who has bandwidth but who does not have the ideal skillset to own the risk
4. Supplement the resources in the impacted area by adding another resource to handle the increased workload

There are other options, but they are variations on one of these themes. So which approach should we take? Well, if risk is the highest constraint in our prioritization—in other words it is the last thing that we are prepared to compromise—then options 2 and 3 are ruled out because they compromise the risk.

Option 1 is acceptable in any situation where schedule is lower than risk in the constraints hierarchy, and option 4 is valid if the resources constraint (and perhaps budget depending on where the resource is coming from) is lower than risk. If risk is at the top of the hierarchy, then obviously everything else is lower, and in that scenario, we would compromise the constraint that is lowest in the hierarchy first to decide on the appropriate approach.

Although we are focused here on organizational risk management, the *ideal* owner of a risk may well be a project level resource. That individual may be closer to the impact area and therefore both better suited (in terms of skills) for the work and more able to spot early warning signs that the risk is going to trigger or see progress made in managing the risk. We explored this in Chapter 6, and this is an acceptable scenario, but there may still need to be a more senior level resource assigned at the portfolio or program level as the owner. This will give the project level resource *responsibility* for the work (they execute the actions) while the more senior resource has *accountability* for the work (they ensure that the actions are executed effectively). The project level resource won't have the visibility at the organizational level to fully assess the risk situation, and as we will see shortly, some of the risk owner's responsibilities cannot be successfully completed without a greater level of understanding of the bigger picture.

The key to success in the allocation of risk owners often comes down to the person or people responsible for that allocation. We need to view the portfolio and program level allocations as related to one another—there is a single pool of resources available, and we need to ensure that the most significant risks are resourced first. This is commonly those risks that impact the portfolio, but there may be some significant program level risks that are considered more important to the organization than some of the lower level portfolio risks. The allocation of resources should ultimately be approved by the portfolio management function, but the identification of appropriate owners and the recommendation of allocations will inevitably require input from all management levels—project, program, and portfolio. It will be tempting at this point to allocate an owner as quickly as possible in order to start the active management of the risk, but it's much more important to ensure that the right owner is identified than it is to assign someone quickly. The work should be treated as a priority, but it can't be rushed.

Risks identified as appropriate for an acceptance strategy still need to be assigned to an owner at this stage even though they are not being actively managed. At a minimum, the risk needs to be monitored to ensure that it is not changing significantly, and in many cases, a contingency plan will also need to be developed. This allocation will come after the active risks have been assigned, but the work is just as important.

One final note on resource assignment—while consistency in the ownership of a risk is always desirable, there will be situations when a risk needs to be shifted to a new owner during management. This may be because of a staff departure, a change in our understanding of how the risk may impact us, or a higher priority that requires the original owner to be re-assigned.

Once the decision on ownership has been made, then the detailed management approach can be determined. The risk analysis has established the overall risk management strategy, but the risk owner needs to determine how to implement that strategy. We split the determination of risk management approach into two phases in order to best utilize the resources we have available to us. Risk analysis is a specialized function that is frequently performed by portfolio or program level resources, and their work allows them to determine the best approach to dealing with the risk. However, those people may be too far removed from the specifics of the unique situation to be able to determine a detailed approach, and that needs to be completed by the risk owner as his or her first piece of work when assigned the risk.

The owner should work with the assumption that their approach needs to remain consistent with the strategy identified during analysis and they are detailing the specific steps that will be undertaken to manage the risk. This should be viewed as a project plan in miniature—the identification of work elements, the time that each will take (effort and duration), the dependencies between the tasks, the regular monitoring and reporting tasks, any additional resources who are required to assist, and so on. This will result in a number of tasks that will need to be added to the portfolio schedule, within a specific program, or within a specific project if that is where the management activities are being executed.

This work should be flagged as risk management related to assist in tracking costs and effort against the risk capacity, ensuring that any future decisions about resourcing assignments understand that these tasks are risk related.

If the estimated work and/or cost of the tasks that form the detailed risk management activities are broadly in line with the estimates developed in risk assessment, then there is no need to identify any variances at this point; they will be captured later on in the process. However, if there is significant variance between the initial assessment and the detailed plan, then this should be flagged so that the risk can be further reviewed and adjustments made if required—they may indicate a problem with the analysis.

In some circumstances, the risk owner may feel that the risk management strategy identified in the analysis phase is not appropriate. One example may be the recommendation to combine transference through outsourcing with mitigation. The owner may feel that there is no suitable vendor to transfer the risk to and therefore wishes to focus solely on the mitigation elements of the work. This is an entirely acceptable approach to managing the risk, but it should require approval by the portfolio management function before it is implemented, regardless of the level of the risk owner, as it represents a variance from the recommended management approach.

Once the detailed management plan is developed, then the risk management process becomes easier, with activities moving into the execution phase—effectively no different from any other project task. The specifics will, of course, vary from risk to risk depending upon the actions that are being taken and the money that is being spent, but broadly speaking, risk management at this point is fully integrated with the work schedules of the various projects/programs/portfolio.

Once the detailed risk management actions have been determined and are being implemented, the focus needs to immediately shift to the development of the contingency plan. This plan details the steps that should be taken if the risk triggers in order to minimize the impact and allow for as rapid and effective a recovery as possible.

The contingency plan has several key elements:

- *Description of the risk trigger.* This is fundamental but is frequently omitted from plans—we need to know whether the risk has triggered in order to know whether we need to implement contingency. Sometimes it's obvious that a risk has triggered (a major failure or key event), but in many cases, there is no dramatic event, just a gradual worsening of the situation. By establishing objective criteria in the plan, we can avoid debate about whether a risk has occurred and simply compare reality with the defined criteria.
- *Steps needed to minimize the damage.* This is work conducted immediately in order to try and prevent the problem from

getting worse and to minimize the impact on portfolio constraints. This may include communication to other areas of the organization that might have exposure to the same risk, stopping of any payments to a vendor who is failing to deliver, realignment of resources to other tasks who would otherwise be stalled until the issue was resolved, etc. At the organizational level, the most important part of this section may well be the process immediately communicating news of the trigger to the appropriate resources. We will look at how that works in the contingency and impact assessment section.

- *Steps needed to recover the situation.* This element of the contingency plan is actually less important at the organizational level than it is for individual projects. When we think about project level contingency, we are focused on finding other ways to meet the project's tangible deliverables. However at the portfolio level, our focus is on achieving the intangible business objectives—an x% revenue increase, a y% cost reduction. That means the steps taken to preserve or recover the business goals may be completely unconnected from the original work that has been derailed, and recovery is not necessary. This isn't always the case, there may be some steps that can be taken to recover the deliverables and the objectives, but this cannot be assumed. For organizational risks, this section of a contingency plan may be excluded in some situations, or may not be implemented in the event that the risk triggers.

- *Approval mechanisms.* Not all of the actions contained within an organizational level contingency plan can be automatically implemented when a risk triggers. Generally speaking, the actions to minimize the damage can be initiated without any additional approvals (although the damage limitation activities themselves may have approval processes in them), but the recovery actions will need to either be handed off to portfolio management for review or will need a portfolio level approval before they are executed. This helps to ensure that effort isn't suddenly refocused inappropriately at a project or program level simply because of a contingency plan—it allows for validation of the recovery while still keeping the damage limitation as swift and efficient as possible.

It's important that the contingency plan is developed early in the risk management processes because it needs to be in place prior to the risk triggering. If the organization is faced with a serious issue caused by a risk becoming real, then the response needs to be swift and decisive—a simple execution of a developed and reviewed damage limitation plan without having to analyze the appropriateness of the response. That said, the contingency plan needs to be a living document—it has to evolve as the risk management activities have impact, as the risk exposure shifts, and as the projects, programs, and portfolio progress. It is vitally important that the contingency plan is accurate and current at all times because if a risk triggers, then the damage limitation activities at least will be executed without further analysis. If the contingency plan is incomplete or out of date, then it may make the situation worse.

In typical project level risk management, ownership of the development of the contingency plan rests with the risk owner. That makes sense on projects because the risk owner is the person closest to the risk and best able to determine the impact. However, it isn't necessarily sufficient at the organizational level. When we consider portfolio and program level risk management, we have already established that there can be separation between the activities that create the risk exposure and potentially trigger the risk and the places where the impact is felt. That's why we created the concept of a portfolio or program level resource being accountable for the risk and working with a project level risk owner who is responsible for the risk. This effectively gives a split in the monitoring role—the project level resource monitors the risk management actions and their effectiveness, the more senior resource monitors the potential impact.

With contingency planning it is the impact of the risk triggering that is driving the response, and so portfolio or program level and project level resources will need to collaborate on the triggers and the damage limitation activities to ensure that actions are both realistic and helpful. Of course, if the risk owner is already a more senior program or portfolio level resource, then the same person can perform both functions.

Once the contingency plan has been completed, it needs to be reviewed and approved at the organizational level. This isn't so much a sign-off as it is a confirmation that the plan accurately reflects the necessary and possible actions based on the current understanding of the risk. This review will be revisited throughout the risk's life cycle as it drives changes into the contingency plan. Frequently this review and approval will be conducted as part of a PMO's control function although it may simply be a peer review by colleagues.

In addition to contingency planning and the execution of specific risk management activities, the risk owner(s) also has to ensure that the risk they own is regularly monitored and if necessary the management actions adjusted. This can be considered a *mini analysis*—it's a regularly scheduled review of the risk in order to determine how it is responding to the work that we are undertaking to manage it. We are obviously looking for evidence that our actions are reducing the chances for the risk to occur and/or the impact if it does happen, and if those improvements are not happening, then we need to decide whether our approach is the right one.

This is a complex situation. We have already discussed how risks are always changing, so we shouldn't assume that things will stay the same, but we need to try and separate the changes that are happening because of the overall risk environment from the changes that are occurring in response to our actions. The monitoring that is carried out by the risk owner will be subjective in many cases—it will be based on skills and experience unless quantifiable metrics are available. The frequency of these mini reviews will be set out in the risk management plan; they may well be weekly for some of the more severe risks and should be no less frequent than monthly, even for passively managed risks where the monitoring is the only conscious action that is taken.

These reviews will seek to update the risk exposure metrics, although this isn't an exercise in measuring numbers. Rather, the frequency of the reviews allows us to look beyond individual changes in each period and consider trends that are likely to be more indicative of the true situation than a simple snapshot that can easily be skewed by a unique set of circumstances. That's not to say that a major shift in the risk environment can be ignored the first time that it appears. It should be reported, but if the event is that significant it will affect more than one risk and will cause organization driven changes to a large portion of the risk list, likely driving fresh risk analysis activities.

This leaves risk owners to focus on the risk specific changes that are occurring, and there are three main drivers for those:

1. *Risk management driven change.* If we have done our risk analysis accurately, then we should see a reduction in one or more of the risk variables (dollar impact, effort impact, likelihood to trigger).
2. *Risk driven change.* This is a change to the risk scenario itself, but at a risk specific level rather than at a wider environment level—that is, a unique characteristic of the risk has shifted. This may be a worsening or an improvement in the exposure or in the likelihood of the risk triggering.
3. *Organizationally driven change.* The progress made in the portfolio or in the project/program components that directly affect the risk may well impact the risk. Completion of a piece of work

that could have been heavily affected if the risk had triggered will lower the exposure for example.

In reality, the risk owners may need to balance a combination of these factors, but this is not the onerous task that it may first appear to be, even if the risk owners are responsible for several risks. This is because they are responsible for executing the risk management activities and so will be familiar with the actions taken and the monitoring that is occurring on a frequent basis, resulting in the ability to observe and react to evolving situations. If we have assigned risk ownership appropriately, the owner will also be working with the variables that drive the risk as part of other responsibilities in the portfolio.

Each time that a risk is reviewed, the risk owners need to determine how risk management should proceed until the next review. There are a number of options:

- Maintain the existing management approach. This should be appropriate in most situations and will reflect indicators that the management activities are having the desired effect in bringing the risk under control.
- Reduce the management approach. This scenario should only occur if the risk exposure has been reduced to the point where we are confident that less active management approaches will not result in the risk exposure returning, or if we feel that risk management is failing and we want to reduce the amount of money and/or effort that we commit to a risk that isn't responding. This may result in changing the strategy to acceptance to reflect the changed situation—the risk is still being managed, unlike in the cease risk management bullet below, but the active management work is reduced or eliminated in favor of passive management. This move should be reviewed and approved at the organizational level.
- Adjust the risk management approach. This may be a change to the detailed steps within the overall strategy identified during risk analysis or a fundamental change in the strategy itself. If the strategy is going to change, then portfolio or program management approval is needed and the risk may need to be resubmitted to detailed risk analysis to understand what has changed and ensure that the measures being taken are still appropriate. This change is often used if the original risk management approach is not working, but it will also be utilized if the risk itself has fundamentally changed because of the environment or a shift in the portfolio or program.

- Cease risk management. This would only be appropriate if the risk has been eliminated, a scenario that usually occurs when the underlying portfolio and/or program elements that are exposed to the risk are completed.
- Trigger the risk. If the trigger events (as we defined earlier in the contingency plan) have occurred, then we need to shift from management to contingency. By definition, the triggering of the risk will also result in the cessation of risk management activities.

Of course, all actions (even maintenance of the status quo) should be documented and tracked in the individual risk summaries and risk management plans.

The execution and monitoring will continue on a regular schedule, with adjustments being made where necessary. Program level risk management will eventually end as the program is completed, although this may not occur for several years on larger programs. There will never be an end to portfolio level risk management as new elements will always be entering the portfolio to replace initiatives that are being completed. We will see a change over time in the specific risk owners and the risks being managed, but the process will be ongoing.

Outputs

The outputs to risk management are simple—the updated risk management plans and risk summaries. The plans are maintained and updated with the detailed risk management actions that we take and provide a consolidated reporting tool for risk status. The individual risk summaries add another level of detail, provide a history of the risk exposure, describe actions taken, and give results of the monitoring. Some of the individual risk summaries will complete their life at this point because the risk has been eliminated, and these documents can then be added to the organizational archive for use as future reference tools. Ultimately, all risk summaries will be added to this archive, but some of them will need to go through contingency and impact assessment first.

11

Process Framework—
Contingency and
Impact Assessment

By the time we reach contingency and impact assessment, we are technically not dealing with a risk anymore. By this point, the risk has triggered, and all uncertainty is removed. Now we need to implement the contingency plan we have developed and carry out an impact assessment to understand what the lasting effect of the risk is. We may also undertake some additional work to attempt to recover from the risk's impact or to initiate alternative work to preserve the portfolio and/or program goals and objectives. The process flow is shown in Figure 11.1 while Table 11.1 summarizes the inputs, process elements, and outputs.

Inputs

The inputs to the contingency and impact assessment process will be the individual risk summaries that provide all of the details about the specific risk that has triggered and the contingency plans that have detailed all of the specific actions that are to be carried out now that the risk has occurred. It's assumed that if our risk management processes to this point have been successful, the volume of items that enter this step will be relatively low, but that may not be the case. There may be many scenarios where the organization has decided to accept the possibility of risks triggering and will deal with them at this point through contingency rather than try to manage

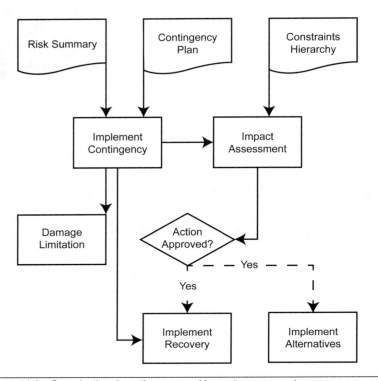

Figure 11.1 Organizational contingency and impact assessment process

Table 11.1 Contingency and impact assessment process summary

Inputs	Process elements	Outputs
Individual risk summaries	Implement damage limitation	Organizational project archive updates
Contingency plans	Conduct impact assessment	Updated portfolio/program risk management plans
	Implement recovery	Updated portfolio/program risk lists
	Implement alternatives	New risk candidates

them proactively. Typically, these will be risk tolerant organizations where risk is low in the organizational constraints hierarchy and where the organizational risk profile is skewed to the right-hand side (high exposure and capacity for impact and low management costs and capacity).

This is an acceptable but high-risk strategy and will usually be confined to young, small organizations, especially in areas where time to market is critical—web and mobile app development start-ups are the obvious example. Organizations that have ignored risk management will find themselves in a similar situation but will likely be less well equipped to deal with the situation! The actions taken in this process can have far-reaching consequences on the portfolio and its initiatives, so it's important that the process elements are comprehensive and robust.

Process Elements

Once a risk has triggered there is a need for swift action to ensure that the situation is brought under control as quickly as possible. During the risk management processes we established the criteria that would determine whether a risk has triggered or not, and we documented those criteria in the contingency plan. That helps to ensure that there is no delay caused by debate about whether a risk has actually triggered.

Not all risks are dramatic, but we know from the Introduction that to meet the definition of a risk there has to be the potential to impact the goals and objectives of the portfolio or program, so these aren't trivial situations either. Therefore we need to minimize the amount of thinking and analysis that goes into the implementation of contingency—the closer that we can make it to the step-by-step mechanical execution of previously defined steps, the more likely we are to be successful.

You will recall that when we defined the contingency plan in Chapter 10 we identified three sets of activities in addition to defining what constituted the trigger events:

1. *Damage limitation*—the steps that we take to prevent the immediate situation from getting worse or to ensure that the additional negative impact is minimized.
2. *Recovery*—the steps that we may take to restore the situation to a *pre-risk trigger* status, or as close to that status as we can reasonably achieve.
3. *Approval*—the sign-off needed to execute either of the steps above. Generally this is only needed for recovery as the damage limitation activities are a minimum of what is needed, and the appropriateness of those tasks will have been established when the contingency plan was finalized.

We selected the risk owner in part because of the owner's ability to closely monitor the risk, so he or she will likely be the first person to identify that a risk has triggered, and therefore the person to initiate the damage limitation steps. As we discussed in the risk management section, when we are dealing with organizational risks, the owner of the risk may be quite distant from the part of the portfolio and/or programs that are going to feel the impact of the risk triggering. This may result in the risk owner being unable to implement directly the damage limitation activities. If that is the case, then the contingency plan will define the process for communicating news of a triggered risk to ensure that the message doesn't get lost or misdirected in a complex organizational hierarchy. This is where a consistently applied partnership between all of the levels involved in project execution can really pay dividends—the presence of established and effective communication links will help to streamline the delivery (and receipt) of the information and minimize the potential for it to get lost.

Damage limitation is focused on dealing with the immediate aftermath of the risk triggering and involves three types of response:

1. *Communication*—advising owners of related risks that a trigger event has occurred. Suppose that a vendor who has multiple deliverables within a portfolio has failed to meet a deadline for delivering a final product. If they are late on one item, or if the deliverable fails to meet quality standards, then this needs to be communicated to the stakeholders of the related items so that they can act accordingly—re-assessing exposure, enhancing contingency plans, implementing more aggressive risk management, etc. In this situation, the damage limitation may be indirect. The risk owner of the triggered risk may be concerned predominantly with ensuring that owners of related risks at the program or portfolio level are aware of the situation, where it is down to the owners of the related risks to determine whether there are any direct actions that they need to take. They may ultimately decide that no further actions are required, but the key damage limitation action is to ensure that the change in potential risk exposure is communicated. The owners of related risks can make that determination and be in the best position to avoid the triggering of a sequence of connected risks.

2. *Dependencies*—managing the impact of the triggered risk on related work items in order to try and minimize the impact on the overall portfolio or program goals. In a project, this would mean we move resources and re-sequence tasks to avoid the problem area from derailing the initiative; in organizational risk management, it may require a more significant change. We may

end up delaying or shelving an entire project in order to protect the overall portfolio objectives. If the risk has jeopardized one project's ability to achieve its goals, then we look elsewhere in the portfolio for that contribution to the goals (remember that at the portfolio level our focus is on achievement of corporate goals, not necessarily on completion of project specific deliverables). These decisions will be made at the highest level of the portfolio management structure in conjunction with the organization's executive leadership, but the discussions will be prompted by the implementation of the damage limitation associated with the triggered risk.

3. *Direct impact*—even if neither of the above categories is required, we have the direct impact of our risk to deal with. Something has happened that has reduced the likelihood of the portfolio and/or a program being able to achieve its goals, and this impact needs to be minimized and stabilized so that a more detailed analysis of the situation can occur. The triggering of a single risk may have impact at multiple levels, and we need to make sure that the focus is at the right level—protecting the portfolio's ability to deliver first, then protecting the program, and finally protecting the interests of the project. This is the exact reverse of the situation that happens in many project driven (rather than portfolio driven) organizations. Even if the contingency plan is clear, we have to ensure that the attitude and focus of the people responding to the triggered risk is at the right level of the organization.

Successful damage limitation requires swift and decisive action, and all three of these categories of actions will need to be implemented in parallel.

In undertaking this work, we have to recognize that our damage limitation activities may themselves impact risk management elsewhere in the portfolio, programs, or projects—increasing or decreasing exposure or diverting resources from management activities. We don't have time to conduct a detailed re-analysis of the risks before we implement our damage limitation, but we do need to make sure that there is communication with affected risk owners so that the impact can be managed.

At the same time that we are implementing the damage limitation activities, an impact assessment needs to be conducted to try and quantify the severity of the problem. When we completed our risk analysis, we estimated the risk exposure in terms of money and time. However, we need to validate or adjust those numbers now that we have a real situation to deal with. Additionally, a single summary financial and schedule figure does not provide a true and complete sense of the situation. Impact assessment needs to consider all of the variables:

- Relevance of the identified recovery steps—are the recovery steps outlined in the contingency plan still appropriate now that the risk has become real—that is, is the recovery plan capable of restoring the organization to the pre-trigger situation or close to it? If the answer is no, then we have to consider what modifications to the planned recovery are necessary, how successful those changes are likely to be, and what the impact of those changes on the rest of the portfolio is likely to be.
- Ability to limit the impact—what is the current and projected success of contingency, and what will that likely mean in terms of the difference between the pre-trigger and post-trigger organizational environment? This difference will help to establish the size and nature of the impact of the triggered risk after the contingency steps have been applied to limit the damage.
- Ability of the organization to withstand the impact—what is the current status of the portfolio and programs and based on that can their goals be achieved without a recovery, or with only a partial recovery? The physical impact of a triggered risk may be the same regardless of when it occurs in the portfolio life cycle, but the larger portfolio environment has to be considered in determining what the significance of that impact is. As a simple example, the impact of a one-week delay a year before the scheduled completion date of an initiative will be different than the impact of a one-week delay just a month before the scheduled completion date.
- Importance of recovery—how important is it to the organization that this risk is recovered from both tangibly and intangibly? This can consider any number of different areas—customers, employee morale, the impact on downstream initiatives, the message it sends to internal and external stakeholders, etc. Other areas of the portfolio may be able to compensate for the impact and ensure that the goals and objectives are preserved, allowing us to avoid diverting resources to recovery activities. However, we have to be aware that this can send a message to stakeholders that the risk *doesn't matter*, which can

have a negative impact on customer perception and employee commitment to risk management.

- Alternate strategies—if we don't try and recover, then what other courses of action are available to us that will deliver equivalent gains? With the focus on preserving the organizational goals and objectives, the basic need is to address any shortfall in those objectives created by the triggered risk. In some scenarios, the most likely (and potentially least risky) way to achieve that is to adjust the constraints of one or more other initiatives within the portfolio, even if they have nothing to do with the triggered risk. This addresses a revenue shortfall caused by the triggered risk by adding a scope element to one project and bringing the delivery date of another forward, even though those projects have nothing to do with the risk that triggered.

The people involved in conducting the impact assessment will include the risk owner, along with project and program managers from affected areas and representatives of portfolio management. As the assessment continues, additional expertise may also be called upon to contribute. For example, if the expansion of scope on a different initiative is considered to be a viable alternative to trying to recover from the direct impact, then the project manager of the initiative that may see a scope expansion will need to be included to estimate the impact on their project.

The impact assessment will ultimately result in a set of recommended steps to recover the situation, to implement alternative solutions that will result in similar benefits, or to take no action. These recommendations then need to be reviewed and approved before they are implemented. The approval process needs to involve management at the portfolio, program, and project levels, although ultimately it is portfolio management that has the final say because the overriding goal is to preserve the ability of the portfolio to achieve its goals and objectives. Program and project management will be responsible for confirming that the proposed steps can be implemented successfully and that they can be accommodated without significant impact to other portfolio risks (there likely will be some project level risk impact, but if this can be absorbed without any portfolio impact, it is acceptable).

Decisions are not always going to be straightforward—we may need to carry out work in order to ensure that we don't lose a major customer or to avoid falling foul of a regulatory body, for example, even if the pure impact analysis suggests that we don't need to take any further steps to recover from the triggered risk. We also have to recognize that some situations may be complex and require us to undertake some preliminary recovery work before we are able to determine whether the implementation of a full recovery plan is appropriate.

If recovery and/or alternative work is approved, then it will move forward and be planned and executed just like any other change to an initiative. The items are included in Figure 11.1 for completeness (with dotted arrows to reflect that they are optional actions), but they are not really part of the risk management process; rather, they are risk driven changes that will be subject to the standard change management process that is used for the portfolio. This work will drive additional risk identification work, and we start the cycle all over again.

Regardless of the outcome of the risk assessment, there will be some recovery work that is necessary in all situations. It may not be captured within a contingency plan, but it is one of the most important tasks to be performed. A triggered risk will impact the people who are working on the impacted areas of the portfolio, programs, and projects, and this impact can be dramatic when a large risk triggers.

A risk event may damage morale and team relationships and lead to finger pointing and the assignment of blame. This impact cannot be ignored. Portfolio and program management are first and foremost leadership functions within the organization, and this is the type of situation where strong leadership is needed. There needs to be time and effort put into helping the team talk through the issues, understand what led to the risk triggering, and understand if there are lessons that can be learned in order to help prevent future problems (making sure that those lessons are captured). Above all, there needs to be an allowance for the time and space that is needed for the team to work through the situation and reengage with the work. This may have a short-term negative impact on the portfolio and its programs, but if that time and space is not provided, then the ongoing impact can be much worse. We can never lose sight of the fact that every element of the portfolio depends on the people that make up our project teams.

This may be simple and straightforward and require nothing more than a few minutes of encouragement in a status meeting for some of the more *ordinary* risks, but if a major risk triggers with significant impact, the implications on teams cannot be underestimated or ignored.

Outputs

Once the damage limitation, impact assessment, and (if appropriate) recovery activities have been completed, the cycle of risk management for these particular risks are complete. The individual risk summary now becomes an update to the organizational project archive, so it can be captured for future reference; we update the appropriate risk management plans and risk lists to close the triggered risk. This is one step often overlooked—with all of the focus on contingency and recovery, the need to capture an accurate record of what occurred is easily missed. The process needs to ensure that there are safeguards in place to capture this information to aid risk management activities on future initiatives.

Finally, we identify any new risk candidates generated by our recovery and/or alternative work, and the cycle of risk management starts all over again.

12

Process Framework— Adjust and Refine

Look back at Figure 7.2 and you will see that we have this adjust and refine process in the center of the cyclical process that we have just detailed in the previous four chapters. We mentioned in Chapter 7 that this reflected the interconnectivity between different elements in the process, and it's now time to look at this function in more detail. We need to treat this process element differently from the others in the process because it is less predictable—it doesn't have consistent inputs and outputs in the same way that the other process elements do. However, in many ways it can be the difference between success and failure in organizational risk management.

The best way to think of the adjust and refine process is as the oversight function for organizational risk management, the element that controls the *how* of the various process elements that form the *what* of organizational risk management. It acts as a management and control function for work that is moving within the organizational risk management process in a nonstandard way and for changes that are being driven from external factors.

There are two main categories of work covered by this process:

1. Triggering variations to the standard processes within organizational risk management
2. Incorporating changes to risk management that are driven by other organizational elements (and in particular other portfolio and/or program management processes)

Before we look at each of these in more detail, let's spend a bit of time exploring why the functions are necessary.

In traditional project risk management, the risk environment is mostly self-contained. The project impacts only a relatively small part of the organization and so is subject to relatively few external influences. As a result, there are unlikely to be many material changes to the risk environment of the project, and if there is a change, then it is probably also driving significant change to other elements of the project—impacting a number of constraints, not just risk. When we move to the program and portfolio levels, the environment becomes far less contained—at the portfolio level we are exposed to any and every one of the risk drivers that affect the organization—and as we saw in Chapter 1, there are a lot of those. Additionally, changes at this level are potential *organization changers*; they can have significant impact on the portfolio and its goals and objectives.

By definition, a standard process will apply the same sequence of activities to each item that passes through that process—and that's the point, a standardized approach that is effective in 99% of the situations that the portfolio handles. The problem with that at the portfolio level is that the items that fall into the 1% are likely to have significant impact on the organization (if they don't have that potential, then they won't be portfolio level risks), and they do need to be treated differently. We need to effectively and efficiently differentiate those organization changers from the normal risk events that happen every day and then act on them appropriately. That's where the adjust and refine process comes in, and that's why it is so vital. If we differentiate the wrong things, then we make matters worse; if we make everything an exception, then we may as well not have a process; and if we fail to identify the items that do need to be treated differently, we are potentially increasing the overall risk exposure unnecessarily.

Let's look at the two different categories in more detail.

Variations from within Risk Management

As you can see from Figure 7.2, the adjust and refine process area has a two-way connection to each of the other four functions within the organizational risk management process. This allows us to connect any two elements of the process simply by going through this adjust and refine element, creating a *short cut* from any point to any other point, as long as the adjust and refine process allows the work item to take that short cut.

Effectively this simply means that any variation to the standard cyclical process has to go through the adjust and refine process before it is allowed to proceed. In some cases these are easy decisions—the identification of a new risk candidate by a risk owner while conducting risk management activities, for example. At a project level, this should probably be considered a failure in our risk identification; at the program or portfolio level, it may well be a frequent occurrence due to the complexity and number of variables. That doesn't mean that we shouldn't look for any process problems that contributed to the risk being missed, but we have to recognize that the environment is so fluid that this will be a regular occurrence. Clearly we want to get the new potential risk into the identification and analysis processes as quickly as possible, so the risk owner simply sends it through the adjust and refine process in order to get the risk confirmed and into analysis as effectively and efficiently as possible.

Other situations may be more complex. If the portfolio manager observes that there seems to be an increasing number of risks coming from a specific area within the portfolio, then the portfolio manager may initiate additional risk analysis of that subset of risks to see if there is an underlying factor that is being missed or misunderstood for example.

Figure 12.1a shows the process flow for internal variations. The process is straightforward and is designed to ensure that the change can be implemented as quickly as possible. It consists of two possible sources of potential variances from the standard process: a request from within the process elements themselves, such as the example of a risk owner who identifies a new risk, and the creation of a variation candidate by a stakeholder that is not directly associated with an existing process step, such as the example of the portfolio manager requesting additional analysis.

In both situations, an analysis step is executed that will review the requested variation to ensure the full impact of the variation has been identified and understood, that the proposed actions are appropriate, and that the work is necessary. Then the work is approved, which will trigger the additional work within the destination process area. The approval shouldn't be seen as overly formal; it is intended to be a checkpoint to ensure that effort isn't wasted on variations that are misguided or irrelevant. It may be nothing more than an initial *sanity check* by the person who will be handling the variation to make sure that it is a good use of their time. It is important to ensure that the approval is objective, approving a request simply because it has come from the portfolio manager is not a good use of resources and potentially also increases the overall risk exposure of the portfolio by diverting resources from where they can have the most benefit. To continue the example from above, if there is a

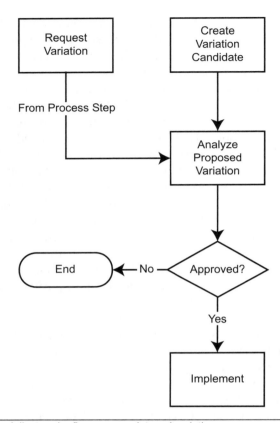

Figure 12.1a Adjust and refine process: internal variation

simple explanation for why one particular area of the portfolio is seeing increased risk exposure, then further analysis is not appropriate and the request should end with the request not being approved.

Externally Driven Variations

Variations from outside of the risk management process are the ones that have the potential to be the most damaging to our control of risks. This is the way that major impacts will hit the portfolio elements—a shift in corporate priorities, changes in the organizational constraints hierarchy, etc. These will have the potential to change the fundamentals of our risk management approach and require us to re-assess the entire suite of risks against the new reality.

Slightly less dramatic, but still a major impact, will be changes driven by events within the project portfolio or an individual program. Suppose that a project is falling behind schedule and actions are needed to correct. If the constraints hierarchy for the portfolio has schedule higher than risk, then we may need to adjust risk management activities in order to free up resources to work on schedule recovery activities (sacrificing risk management and accepting a higher level of risk in order to try to preserve the time frame). This might result in a shift from active to passive risk management for a subset of the risks or fewer resources conducting analysis.

The process for external variations is shown in Figure 12.1b and is even simpler than the one for internal variations.

With changes driven from outside of the risk management process there is no opportunity to decide whether the variation will be accepted—it's real, and we need to handle it as quickly and effectively as possible. However, we need to conduct an impact assessment to make sure that we understand the extent of the change that we are facing, the steps that we need to take, and any further work that needs to be carried out. Changes driven from outside of the risk management process tend to be fairly dramatic and have the potential to cause significant disruption to the execution of risk management. It's important to ensure that the impact

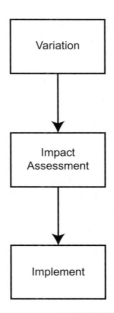

Figure 12.1b   Adjust and refine process: external variation

assessment is objective, but also that it considers the impact on the people involved.

Suppose for example that the organization decides to make a significant shift in the portfolio partway through the execution cycle. A major new program is approved, and a number of other initiatives are cancelled. This will drive a number of risk management changes through the adjust and refine process:

- Risk identification work for the new program. This may require additional resources to be made available for the identification work and may form a knock on resource need in the risk analysis process. By the time these reach management, there should be assigned resources, and the increased demand will likely be capable of being absorbed.
- Closure of risk management for the initiatives that are being cancelled. This will involve a number of different tasks—updating risk management plans and risk lists and archiving of documents, for example.
- Modification of remaining risks. Those existing risks that are not directly affected by the closure of projects are still going to feel some impact. At the least their relative priority will change, which may drive changes in the way that they are managed, and there may also be changes in exposure and management approach caused by the way that the risks relate to the new program.

It's relatively straightforward to analyze the impact on tasks, and while the changes may be dramatic, the work can be defined and scoped. However, there will also be an impact on the people who need to execute these processes. Some people will find that the work that they have been focused on for a considerable period of time has suddenly been cancelled, they may fear for their jobs, they may have to get used to a different role within the organization, etc. The adjust and refine process needs to consider the human impact of these changes both from a practical standpoint (loss of productivity and increase in error rates) and a leadership standpoint (allowing people time to come to terms with the changes and allowing increased flexibility in the approach to work while people adjust to the new reality).

13

Portfolio Level Risk Management

In the previous five chapters, we detailed each of the sections in the organizational risk management process framework and looked at each step in the process, as well as the specific process elements involved in each of those steps. Now we are going to start looking at how each of the different levels of organizational project execution interacts and applies those processes and at some of the unique considerations that occur at each level. We'll start at the top of the pyramid with portfolio level risk management.

Portfolio Risk Management in Context

Let's start by considering the risk environment. Figure 13.1 shows a high-level view of how portfolio risk management fits into the overall picture.

Looking at the diagram, you can see that all of the relationships are two way. This reflects the fact that risk management is a cyclical activity— risk identification and analysis activities drive the organizational profile, which in turn drives management activity that generates updates to the profile, etc.

We must also be careful not to treat portfolio risk management in isolation from other portfolio management activities. Successful portfolio execution depends on comprehensive, consistent risk management, but it

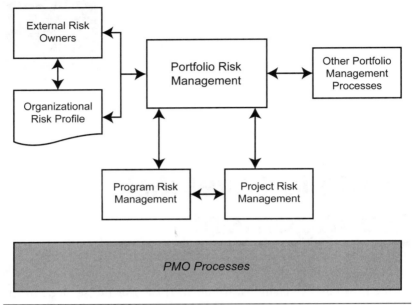

Figure 13.1   Portfolio risk management environment

also requires the same approach to managing quality, budget, schedule, and scope.

The diagram shows program and project risk management processes directly connecting to the portfolio risk management, rather than having a hierarchical structure of going through programs to get to projects. This is a deliberate approach that recognizes that not all projects will be part of programs, but also that the overall approach is about being as effective as possible, treating all of our risk team members as equal rather than imposing arbitrary (and unnecessary) organizational hierarchies.

The Scope of Portfolio Risk Management

Before we start considering the details of portfolio risk management, we need to establish when it begins and ends, and that means examining when portfolio management as an overall concept begins and ends.

Portfolio management is not simply *high-level* project management; rather, it is a strategic function that, as we noted before, is concerned with the organization's ability to deliver on its goals and objectives. I've already touched on this earlier in the book, but effective portfolio management needs to span these items:

- Idea generation—the capturing of ideas from within the organization that may eventually become initiatives in the portfolio. This will cover everything from someone thinking that an aspect of the organization's operations could be more efficient to product managers developing the roadmap for their products.
- Review and enhancement—the consideration of the ideas generated above with a view to improving them and discarding ideas that won't work or don't align with the organization's goals. This will start with simple peer review of ideas and will evolve through teams working to improve those ideas and on to formal business case analysis for the ideas that make it that far.
- Selection and approval—the traditional process of allocating available budgets across the initiatives that are ultimately approved and become part of the formal portfolio. We typically think of this as an annual process, but it doesn't have to be.
- Execution—the mainstream elements of portfolio management that is concerned with managing the execution of the various initiatives that collectively form the portfolio.
- Benefits realization—ensuring that the benefits identified in the business cases are actually achieved by the completed initiatives. There may have been adjustments made to the numbers between approval and delivery, but this is ultimately where the success is measured from a portfolio perspective. It is each initiative *pulling its weight* against the expected contribution.

Your organization may not have as broad a definition of portfolio management, which doesn't mean that you can't be successful, but I firmly believe that if portfolio management is restricted to a narrower focus, it will be less effective, so we are going to treat this as our scope.

If this is the scope of portfolio management, then this also has to be the scope of portfolio risk management. The processes that we outlined in the previous five chapters begin to be applied at the idea generation level and continue through benefits realization. That can be a tough concept to get your head around, especially when it comes to early stages in the portfolio management process, but an example may help to illustrate the point.

An obvious early stage risk might be that we spend a lot of time and energy reviewing and improving ideas that are not aligned with our corporate priorities. If we are focused on consolidating in existing markets in order to survive a tough economic climate, then we don't want to spend a lot of time and energy investigating market expansion ideas or new product development concepts. However, we also don't want to pass up any easy opportunities to gain a competitive advantage by ignoring a good idea.

This gives us two portfolio level risks for the idea generation stage of a portfolio:

1. We invest too much time and effort in assessing ideas that are misaligned with our corporate priorities.
2. We miss out on an opportunity to gain significant competitive advantage in our market by not pursuing an idea that was generated within the organization.

The advantage at this early stage of the portfolio is that there can be no confusion about whether these are program or project level risks—they clearly belong at the portfolio level because there aren't any programs or projects yet, or at least none that relate to this specific element. These risks can both be passed through the organizational risk assessment process. Depending on the organizational goals and objectives, organizational risk profile, and the portfolio constraints hierarchy (among other considerations), we will determine how the risks will be managed. If the organization is absolutely determined to pursue a consolidation strategy and will not consider anything else, then we need to aggressively manage risk #1, maybe even work to eliminate it, while accepting risk #2. In fact, risk #2 may not even be viewed as a risk in this situation because the focus of the organization isn't on improving our competitive advantage.

At the other end of the timeline, benefits realization can provide similar challenges to people who are more familiar with project style risk management. In a project sense, the risks have been dealt with by this point because the project has met its deliverables, but from a portfolio standpoint, this is one of the most crucial phases. Risks that trigger here can have a profound impact on the ability of the portfolio to deliver.

Remember back to one of my recurring points—portfolio management is about delivering the goals and objectives of the organization. The projects and programs that make up the portfolio provide tangible deliverables that contribute to the achievement of those corporate priorities—a new product is delivered to market that increases revenue and market share or a new system is deployed that delivers lower operational costs through greater automation.

Benefits realization is the point in the portfolio where we connect those tangible deliverables with the benefits that they are expected to deliver—the amount of revenue growth, the percentage market share gained, the annualized operational savings, etc. In a general sense the risks here are obvious—we fail to achieve the projected goals, we only achieve some of them, or we achieve the goals but create problems elsewhere.

These are also project and program level risks—we should be taking steps during execution to ensure that the deliverables of each initiative have the best possible chances of meeting the goals. However, there are also portfolio risks here that are separate and unique. At a somewhat generalized level, here's an example of how risks at different levels all impact benefits realization for a new product that is designed to gain market share and revenue growth:

1. Project level. The features don't meet the needs of customers and the product is rejected by the market. We manage this risk by working with existing and potential customers to understand the features that they are asking for, seeing what competitors are doing, carrying out focus groups, and so on.
2. Program level. The customers who adopt the first iteration of the product have demands for additional features and functionality that detracts from the ability to deliver on later phases of the program and delays functionality that can generate more revenue. In addition to the risk management that we carry out at the project level, we may manage this risk by being careful about which customers we select for early phase or pilot deployments or giving customers favorable terms in exchange for an understanding that they may have to wait for custom features.
3. Portfolio level. There is insufficient market demand for the product, a competitor steals market share through aggressive pricing or a rival product, we incorrectly estimated the willingness to pay the price that we have set, or the sales process is taking longer than expected. We try to manage these risks by obtaining early commitments from customers in exchange for favorable terms, showing early development versions of the product to potential customers, and ensuring that sales teams are supported and trained.

As you can see, there are three different sets of risks with a common impact. The particular challenge that we face when we reach the benefits realization phase of a particular portfolio element is that the project is already complete, so we can't leverage that to help manage the risk. The program has moved on to the next phase and is focused on the other projects of which it is comprised. If we try to leverage that, then we will create additional risks and issues in those downstream projects.

That leaves the portfolio somewhat underpowered by the time we reach benefits realization. The portfolio generally acts through programs and particularly projects, as we explored when we defined our process. By the time we reach benefits realization, we need to have completed most of our risk management (because the projects are complete and the program is focused elsewhere); now we have to wait to see whether the goals will actually be achieved. If we are lucky, then we have sufficient commitments that we can be confident the goal will be easily achieved. Perhaps preorders are even high enough that the goal has already been achieved, but in most scenarios, there is still a degree of uncertainty. There may still be some things that we can do to manage the risk for this specific product—price promotions to gain market share, for example. However, they often involve compromise—revenue for market share, ongoing maintenance revenue for upfront sales revenue, etc. By this point, portfolio risk management should be taking a higher level, more strategic approach.

When we looked at the organizational risk management processes, we established trigger events in our contingency planning and triggered the risk if those conditions were met. Portfolio risk management needs to establish those triggers for the benefits realization of each initiative that forms part of the portfolio. We need to be looking at trends, so we can predict whether we believe the project will achieve its goals (we can't wait until the end of the reporting period to see whether the goals are met because then it will be too late to address any shortfalls). If not, then we trigger the risk and implement our contingencies. These may include:

- Adjusting subsequent initiatives within a program to address the shortfall (changing the scope or release date) to try to address a product shortcoming or to add more value to a product that is perceived as more expensive
- Creating additional initiatives within the portfolio—brand new, unplanned projects that will deliver product features, improved deployment, and implementation processes that will aid the product
- Driving operational work—training and support for sales teams and implementation engineers to help them with the sales and deployment cycle
- Adjusting other portfolio elements that have nothing to do with the product that is missing its targets—e.g., accelerating the deployment of other revenue generating and/or market expansion initiatives to increase their contribution to portfolio goals or adding features to other product initiatives to allow them to generate more revenue or gain market share.

The last of these is actually likely to be the most appropriate strategy for a well-managed portfolio, although many organizations will focus on the first three. There is nothing inherently wrong with looking for solutions that are related to the problem area, and those solutions may be easier to implement, but we need to make sure that portfolio risk management is taking an organization-wide approach to managing risks.

That means recognizing that the measure of success comes from achieving the goals and objectives of the portfolio, not necessarily the goals and objectives of every portfolio constituent.

Resourcing Portfolio Risk Management

In project management, the resourcing model is generally fairly clear—we have a project manager and a team of people. The individuals that make up the team may change as the initiative progresses, but when we talk about the project team, most people know what we are discussing.

Portfolio management resourcing is a little bit more of a complex concept. Reading between the lines of some of the details of the organizational risk management process in Chapter 10 suggests that I was referring to a team of people, but what does that really mean? Before we look directly at risk, let's take a step back and look at some models for a portfolio management organization.

Portfolio management is a control function, not a day-to-day management function. The specifics of the role may vary from one organization to another, but it is concerned with maintaining alignment between the objectives of the work and the results—resolving issues that come up and creating an environment that is conducive to success. Clearly, this is a significant piece of work, but portfolio management doesn't need a team of a hundred people!

In many organizations, portfolio management is not a full-time job, even for the *portfolio manager* (or whatever title is given to the person ultimately accountable for the success of the portfolio). Instead, the portfolio management function is grouped together with the PMO, and the terms are used nearly interchangeably. This situation is reinforced by the numerous different PMO models that exist, with different areas of responsibility, reporting lines, and staffing models. It's easy to see why it is convenient to align portfolio management and the PMO.

However, the two functions are very different, and we are going to consider them that way. I have no problem whatsoever with the idea that portfolio management resources are organizationally aligned with the

PMO, and I fully accept that individuals may have a role to play in portfolio management and a role to play within a PMO infrastructure, but those roles are separate and need to be treated that way. We'll look at some of the complexities that can be caused by inconsistent reporting lines when we get to the chapter on the PMO.

For our purposes let's use the following definitions for PMO related work and portfolio management related work:

- PMO work is concerned with the support functions associated with the execution of projects, programs, and portfolios. This might include process governance and audit training, process improvement, consolidation of reporting and administrative services (plan maintenance, financial reporting).
- Portfolio management is the strategic level function concerned with the tasks that directly impact the success of the portfolio or its constituent parts. This will include monitoring progress against plan, analyzing variances, resolving issues, resource management, change management, budgetary control, and of course, risk management.

So any time that an individual is working on one of the elements in that second bullet point, they are performing a portfolio management function, regardless of their nominal job title or reporting lines.

Now let's bring this back to portfolio risk management in particular. There are two types of experts that we will need:

1. General experts in portfolio and risk management who can assess the majority of risks, make decisions around the management approach, act as or work with risk owners during the management process, and/or conduct impact assessments. These people will likely be assigned to portfolio management functions on a permanent or at least semi-permanent basis for a percentage of their work—say 80% PMO and 20% portfolio management. In larger organizations, they may even be permanent portfolio management resources (although they will still likely have other portfolio management responsibilities beyond risk management unless the organization is extremely large).

2. Specialized experts who are called on to assist with specific situations—for example, to help with qualitative and/or quantitative analysis in an area where specialized knowledge is required in order to be successful, or product-focused resources who assist in the identification of triggered risks in the benefits

realization process described above. These individuals are more like project resources who are assigned to the portfolio for a specific task and then have no further commitment and return to their main roles.

In addition, we also need a decision-making process for the various reviews and approvals that are required, and this should be as consistent as possible and avoid a need for multiple levels of approval. My preference is always to have the portfolio manager no more than one level away from any approval—that is, that all approvals are carried out either by the portfolio manager or by an individual who reports directly to the portfolio manager for their portfolio work.

The constant challenge with resources who aren't dedicated to a particular team is getting them to see their *secondary* role (which portfolio management will frequently be) as important—often it gets given a lower priority than the primary tasks controlled by their line manager. This can be managed in part using the organizational constraints hierarchy that will define the relative importance of risk (and by extension risk management) against other priorities. If risk is at the top of our constraints hierarchy, then we shouldn't be losing resources to non-risk related tasks. If risk is at the bottom of the organizational constraints hierarchy, then we may well have to wait for resources or accept a lower resourcing level.

However, a lot of the work on resourcing will be associated with the portfolio management culture within the organization. At a fundamental level, if the organization isn't actively committed to portfolio management and ensuring that it is positioned as the key to successful change initiatives, then securing resources will always be a challenge. On the other hand, if the organization has a Chief Project Officer or similar role that implies a project based culture that permeates the entire organization, then there will be an expectation that people will be required to assist at various times.

Managing Portfolio Risk Changes

The organizational risk management framework that we defined in the previous chapters is intended to apply equally well at the portfolio and program levels, and at that lowest level—the individual risk—that is the case. Where we run into additional challenges with portfolio level risks is the far more significant breadth of risk exposure—there are more and larger areas that can see things go wrong. It's not an exaggeration to suggest that every one of the external risks that we identified in Chapter 1 has the potential to impact the portfolio significantly. When we looked at the organizational risk profile, we talked about it only ever being a *snapshot* of a risk environment that was constantly evolving and changing. The portfolio affects, and is affected by, a large percentage of the organization and as a result is exposed to a lot of those shifts and changes in the risk environment.

This in turn will lead to constant shifts in the risk exposure within the portfolio—the situation that we assessed one day to determine our risk management approach may be completely different the next day and can call into question the entire strategy for managing that risk. At the same time, if we respond every time that something changes in the portfolio's risk environment, then we will spend all of our time assessing and re-assessing risks and never achieve anything. Note that we are only talking about external factors here—things that impact the portfolio from outside, not changes in individual risks or portfolio risk tolerance.

This brings in to play one of the most difficult, and important, elements of portfolio risk management—the ability to control change. Let's start with the concept before we look at specifics.

I like to think of this concept as being like a shock absorber in a car. The name of the part says it all—it absorbs shocks in the form of unevenness in the road. If you are riding in the car, you are still aware of some of the significant bumps, but the minor ones are smoothed out, leaving you with a more comfortable ride. In portfolio risk management, the same thing applies. The shifts in the risk environment are the equivalent of the bumps in the road, and there needs to be a risk management equivalent to the shock absorber that can deal with the minor inconveniences and ensure that only the major hurdles impact the team members who represent our car's occupants. So how do we do that?

In many ways this function is itself about risk management—assessing which of the *bumps* can be absorbed and which ones need to drive change into the portfolio is not an exact science and requires a combination of skill and good judgment. This needs to start with having mechanisms in place to ensure that the portfolio is made aware of any shifts in the risk profile in a timely manner.

As we looked at when we developed the risk profile, every risk is given an owner, so there is no excuse for a change not to be recognized from that source—either a triggered risk or a major change in impact and/or a likelihood to occur. The portfolio manager is the overall risk owner for the portfolio parts of the risk profile and is accountable for ensuring that they have strong communication links with all risk owners in order to ensure that changes are quickly communicated, something that should be reinforced with established processes for facilitating that communication.

For changes that originate outside of the portfolio, we rely on strong organizational risk management and communication links. We discussed in Chapter 6 on the organizational risk profile that every risk needed an owner; that individual and the portfolio manager should be communicating frequently to ensure that the potential impacts are understood and monitored. This is simply a case of extending the organizational risk management model beyond the portfolio that we are focused on here and into more operationally focused elements of the organization.

Once there is awareness of the change, we need to assess the impact of what has happened in terms of the portfolio—the key *shock absorption* function. Not every change in the risk environment needs to impact the portfolio—it may be a temporary *blip* or it may be a minor impact (at the portfolio level), in which case it can be absorbed within the existing risk management approach and simply be considered as part of the overall risk environment in future risk assessments. On the other hand, it may require a re-assessment or a change in the approach to some risks, or in some cases, it may need a wholesale replanning. From a process standpoint, the approach can borrow from organizational risk analysis, and if a change is deemed impactful, then it can be considered a variation and enters the process shown in Figure 12.1b.

However, the shock absorption itself has to happen first—absorbing changes where possible and minimizing others so that the key work of portfolio execution can continue. This needs to be a portfolio management driven decision-making process. The ultimate analysis may be conducted by general or specialist experts (see resourcing section above), but the responsibility for determining how much of the shock will be absorbed rests with the portfolio manager.

The first decision that the portfolio manager needs to make is whether to conduct any analysis on the impact of the risk at all—in other words, to decide whether the risk is going to be completely absorbed. This decision will be driven by the answers to two key questions:

1. Do we have the ability to handle the specific increased risk exposure that the change represents without there being a significant negative impact on the portfolio?
2. Are we prepared to commit the required amount of our risk capacity to this change?

To put those questions in the context of the risk management profile that we developed in Chapter 5, we are first establishing whether there is sufficient risk capacity to absorb the increased exposure, and secondly deciding whether we are prepared to sacrifice some of that excess capacity to avoid any tangible impact from this change in circumstances. If the answer to both questions is yes, then we can absorb the change and simply note it for future risk analysis. If the answer to either question is no, or if we aren't sure, then we have to consider conducting an analysis of the change.

Of course, that analysis is itself an impact on the portfolio in that it takes resources away from other activities, which will inevitably have some impact, so we have to be sure that the analysis is appropriate. This will be driven in large part by the risk tolerance, represented by the organizational risk hierarchy—the lower that risk is in that hierarchy of constraints, then the greater the risk tolerance (we are prepared to accept more risk in order to preserve the constraints that are higher in the hierarchy). That will mean we are less likely to conduct the analysis or will reduce the depth of the analysis in scenarios of higher risk tolerance, and we absorb more of the change. On the other hand, if control of risk is highest in the constraints hierarchy, then there is less room for shock absorption, and we need to conduct a more thorough analysis and potentially modify our risk management approaches for some risks in order to maintain control of the overall portfolio's risk exposure.

Note that this is always something that should be viewed from an overall portfolio risk standpoint. We aren't managing each risk individually; we are making sure that our overall portfolio risk exposure remains within acceptable bounds. That leads on to the next portfolio specific element, strategic risk management.

Strategic Portfolio Risk Management

The organizational risk management process is designed to allow for efficient, effective risk management of each of the risks that may impact the portfolio. This is vitally important. Even though the portfolio is the highest level of the project execution approach, we still need to drill down to individual risks in order to ensure that there is adequate control over the things that can derail the portfolio—the *devil is in the details* if you will. However, we also have to ensure that the overall risk management process is managed in a strategic manner, that the collective risk management effort and progress is acceptable and appropriate.

This is considered to some degree when we execute our organizational risk management process. We review the organizational constraints hierarchy to ensure that the relative importance of risks is considered and that the management of risks is adjusted accordingly—that we don't divert resources from higher priority items, that we don't aggressively manage risks that have acceptable impacts, etc.

Now we have to apply a level of strategic monitoring and control to the process that is focused on the bigger picture. This is the portfolio manager's responsibility and will be carried out in parallel with the management of the individual risks. Strategic portfolio risk management is concerned with the following aspects:

- Reserve utilization. Reviewing the amount of contingency reserve and management reserve that has been utilized relative to the overall progress on the portfolio. The desire is to complete the portfolio with all of the management and contingency reserves exhausted. If the projection is that we will run out of reserves before the portfolio is completed, then we may need to become more aggressive in our risk management approach. If the projection is that there will be substantial reserves remaining, then we may want to take a more passive overall risk management approach. In reality, we may never *complete* the portfolio because new initiatives will be added, but the run rate projections will still be valid based on the constituents of the portfolio at any given time.

- Progress against goals and objectives. As already discussed, the portfolio's purpose is to ensure that the overall goals and objectives for the portfolio are achieved. If we are currently projecting a variance (either exceeding or falling short), then it may be appropriate to adjust risk management to try and bring the portfolio back on track. The specific actions will be complex in this scenario and will require an understanding of what is driving the variances to determine whether we need to be more aggressive or more passive.

- Performance against specific constraints. If we have variances in a constraint that is higher up in the hierarchy than risk management, then we may need to adjust risk management in order to help bring that constraint back into line. While we would typically think of this as sacrificing risk management to protect a more important constraint, a positive variance may also allow us to make adjustments. For example, if cost is the highest priority and we are projecting that the portfolio will deliver under budget, then we may be able to invest in more risk management and still keep costs on target (coming in under budget is not necessarily a good thing, as the money has already been assigned to the initiative and should be invested as effectively as possible).

- Risk capacity utilization. The run rate of the consumption of risk management resources (money and effort) may require adjustments to be made. If our investment in risk management is likely to see us run out of capacity prior to the completion of the portfolio (based on the current portfolio makeup), then we need to either scale back on the management and adopt a more passive approach, and/or increase our risk management capacity by making more budget available—assigning more resources to risk management and moving them off of other tasks. On the other hand, if we are not consuming the capacity at the projected rate, then we may be able to free up some budget or resources.

Of course these decisions are not black and white. Just because current run rates are running higher than expected, that doesn't mean that things won't improve as the portfolio continues and risk management starts to have impact. The statistics will provide the portfolio manager with a quantitative analysis of the situation, but as we saw with risk assessment, we need qualitative analysis as well, and that's where good judgment comes in.

Decisions always need to consider the constraints hierarchy—the position of risk in that hierarchy should be the major determinant of the actions that we take, always being prepared to compromise something lower to protect risk but never compromising something higher. At the same time, the portfolio manager needs to resist the temptation to build up a *safety margin* in the higher priority constraints. For example, if schedule is the most important constraint, then it can be tempting to try and stay 10% ahead of plan in order to build up a buffer in the event of a future delay. That's not automatically a bad thing, but if it compromises quality, then it's certainly not good—a customer would rather have the product that they paid for on time than a sub-par product a month early.

This of course means that portfolio risk management cannot be considered in isolation, but as part of the larger management of the portfolio. This sounds obvious, and it should be, but it can be easy to focus on one aspect of the portfolio and ignore the fact that there are many elements acting together in a complex ecosystem that makes up the entire portfolio. The good news for organizational risk management at least is that it forces focus on all of the constraints because the measure of ultimate success is whether the organization achieves its goals. That won't happen unless there is a balance between all of the constraints.

Strategic portfolio risk management is not limited to the risks that impact the portfolio level; it needs to be an all-encompassing approach that also considers program and project level risks, and this is where life can get complex. The solutions to the challenges identified above may not be as simple as to increase or decrease the amount of risk management that is performed. There may be a need to make decisions that put some risks above others. This effectively allows some risks to go unmanaged in order to protect the management of others. In some cases this will simply be an adjustment to the level of risk prioritization that gets actively managed (i.e., moving risks from active management to passive management), but in other cases it will mean sacrificing projects and potentially programs in order to protect the portfolio. An example will help to illustrate the point.

Suppose that risk is the lowest of the constraints in the risk hierarchy. We cannot sacrifice any of the other constraints in order to protect risk management, so if we find ourselves with a risk management run rate (money and/or effort) that is too high, then our only option is to cut back on the amount of risk management we perform. We have analyzed the risks and are concerned that if we just lift the threshold for where passive organizational risk management ends and active organizational risk management begins, then we will be increasing the risk exposure in too many places throughout the portfolio. Similarly, we don't feel that there are any specific organizational level risks that can have their risk management cut back without increasing risk exposure to the point where there are serious risks of failing to achieve the portfolio goals. We know that this is an option if there is no other choice because risk is the lowest of the constraints, but we need to try and find a better way.

If we start to look lower in the portfolio hierarchy, there are a lot of individual projects where risk management is being carried out, and those costs ultimately roll up to impact the portfolio's risk capacity—as we discussed when we built our organizational risk profile. Therefore, if we can reduce some of the costs at that level, then we free up some resources for organizational risk management at the portfolio level. To do this we need to try and find a project, or combination of projects, where the overall contribution to the success of the portfolio is relatively small but where the resources spent on risk management are sufficient that they can have an impact if they are reallocated. If we then reallocate some (or all) of the resources currently being invested in risk management, we are maximizing the benefit (better organizational risk management) while minimizing the cost (greater chance of a part of the portfolio failing).

In this scenario, we are accepting the *lesser of two evils*—the potential for a failed project caused by inadequate risk management is being consciously accepted in order to try and reduce the risk of a larger failure at the portfolio level. Other considerations also need to go into the decision-making process:

- Dependencies between the sacrificed project and other initiatives—will a failure cause additional downstream problems that may themselves impact the portfolio's ability to deliver?
- Program level connections—will a failure have an impact on future elements of a program that are not yet part of the portfolio?
- Organizational issues—will a failure have an effect on the organization's operations or the ability to launch future products and services?
- Customer issues—will a failure adversely impact relationships with one or more customers?
- Regulatory issues—will a failure cause a potential breach in the organization's compliance?

The answers to these should be simple to determine because they should all have been identified during the project level risk analysis work. The key is to ensure that they are considered as part of the decision-making process of whether to sacrifice some project level risk management to try and preserve the portfolio.

This is where the portfolio manager has to demonstrate a true strategic leadership capability—recognizing that they need to act for the overall good of the organization, even if that sometimes means making decisions that may not support achieving the goals and objectives of the current portfolio. Decisions that protect greater future success at the cost of lesser short-term success may well be both acceptable and desired.

This focus on the ability of the portfolio to achieve its business goals, and the consideration of more than simply those goals, has one other complexity to consider. Because portfolio management is a strategic, organizational level role, it needs to focus on more than simply the portfolio as it exists today. Organizations will set goals and objectives for a year or a quarter, but those should be stepping stones to a longer term plan for growing the business in a number of different ways.

That means that portfolio risk management (and all other portfolio management functions) needs to consider that long-term view. In the same way that we can't make decisions about risks without considering customer impact, regulatory impact, etc., we can't make decisions about the portfolio today without considering what it will do to the portfolio in the future. This longer term view should be used as a form of *checks and balances* for the decisions that we make in strategic portfolio risk management—a validation that the approach that we are taking is aligned not just with the current goals and objectives, but also with the longer term strategic priorities.

This might be harder than it seems looking solely at the portfolio. If an organization uses an annual planning cycle to feed new initiatives into the portfolio, then we may not have an accurate view of what the portfolio might look like beyond the end of the current year. If there is a more ongoing process for approving initiatives, then there will be greater visibility into upcoming approved initiatives, but this will still be short-term relative to the several years of planned organizational growth. Programs offer a better insight into future priorities, as at least some of them will likely be spread over a longer period, but this can also be misleading. If only one program currently extends more than twelve months out, then that doesn't mean that is the only priority a year from now, it simply means that the other work hasn't been defined or approved yet.

This is where the role of portfolio manager as organizational leader becomes crucially important. The portfolio manager needs to have visibility into the medium- to long-term business plan to ensure decisions made today not only help achieve the current goals and objectives, but that they are also consistent with the longer-term organizational growth plan.

If those decisions appear to be at odds with one another, then there may be a need to identify a better approach that is less optimal today but a better overall solution for the company's plans. At the least, there should be a discussion between the portfolio manager and the leadership team that forms the portfolio level stakeholders.

The clear implication of this is that the portfolio manager has to be an organizational leader. While they may not have the formal title of a C-level executive, they have to be considered as a senior leader within the organization—to be effective they need the freedom to make the decisions described in this chapter and that clearly requires a considerable amount of organizational authority and autonomy.

14

Program Level Risk Management

So far we have looked at the organizational risk management process and considered some of the unique challenges that portfolio management faces in applying that process. Now it's time to turn our attention to program level risk management, and things can be difficult in this area. Not only is the program positioned between the portfolio and the projects where it has to deal with issues and influences from above and below, it is dealing with the complexity of time that imposes additional restrictions on our ability to control the risk.

A quick refresher on the basics of a program is worthwhile before we go much further. A program is a grouping of projects with a common purpose that is designed to deliver greater benefits than can be achieved from managing those initiatives on a stand-alone basis. Clearly, program level risk management is one of those benefits. As an example, think about what's involved in hosting an Olympic Games. It's not simply two weeks of sporting events; there is construction, logistics, security, communications, transportation, maintenance, and advertising. It will start literally years before the opening ceremony and will continue for months after the closing ceremony.

By grouping all of these different projects under the umbrella of the Olympics program, we can leverage an overall program level infrastructure—centralized finance and administration for example—to make the execution of each project easier, to make the program more efficient through economies of scale, and to provide a little (or a lot) more leverage when trying to get things done.

In a corporate environment, few programs will be the size of the Olympics, but the concept remains the same.

Program Risk Management in Context

Figure 14.1 shows a high-level view of how program risk management fits into the overall picture. You can immediately see how many relationships there are to deal with here. The constituent projects will roll up to the program, and there will be two-way impacts between these levels, as we would expect. By definition, the program is also part of the overall portfolio and will be impacted by all of the items that we looked at in the previous chapter. At the same time, the program may be driving risks into the portfolio, and so we have a two-way connection here too.

Of course, program risk management has to also consider the other program management processes and it will impact and be impacted by those in a similar way to the portfolio. We also have the potential for

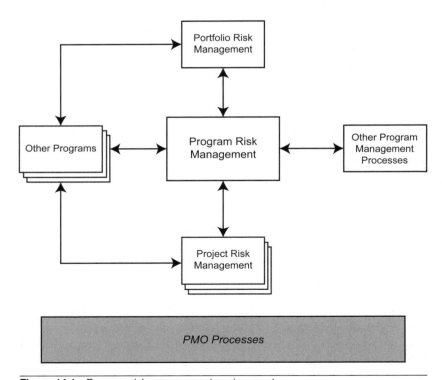

Figure 14.1   Program risk management environment

programs to impact one another—just because the programs are addressing different elements of the business doesn't mean that there can't be risk related connections between them.

There is also one other significant complexity that we have to consider at the program level, and that's the concept of time. When we talked about the portfolio, we were talking about the projects that were underway at any given moment in time. Many organizations will take an annual view to the portfolio—a once-a-year planning cycle that approves the projects that will be executed for the year, and the portfolio will evolve as new projects are started and others completed. This window will be extended when we consider project proposals and benefits realization, but the portfolio is still a relatively immediate snapshot, albeit with the potential to impact long-term goals. As we discussed at the end of the last chapter, there is also a need for portfolio risk management to be forward thinking and consider the longer term organizational priorities. With program management, the concept of time is much more integral to the process.

Think again about the Olympics example. After the Olympics have been awarded, there will be a planning of all of the different initiatives, the dependencies between them, the deadlines and key milestones, and the checkpoints to formally review progress. Most of the work will be scheduled back from the date of the Olympics themselves (because the date cannot shift), and many of the items will be several years into the future. As a result, some initiatives will remain as nothing more than entries on a plan for many months but will still have dependencies with work that is starting immediately, creating a connection that can lead to multiple and severe risk related impacts.

The Scope of Program Risk Management

With program management being influenced by, and influencing, so many other areas, it's important to accurately define the scope. This is relatively straightforward within an individual program—the scope of the program defines the boundaries of risk management—but it becomes more difficult when we consider the program as part of a portfolio and as a consolidation of projects.

The organizational risk management process that we looked at in Chapters 7 to 12 still applies; that cyclical approach will work just as well at the program level as it does at the portfolio level. Program level risk management is still a strategic function—we have to retain a focus on achieving the overall goals and objectives rather than the specific deliverables, just like we have to with portfolio risk management, and many of the considerations that we looked at in the previous chapter on portfolio management will still apply at the program level. Where program level risk management differs is the scope of the strategic focus. With portfolio risk management, we were concerned with the organizational objectives; here we are focused with the program objectives.

That should still be defined in terms of business benefits—the contribution that the program will make to increased revenue, profitability, market share, etc., but there may be three categories of contribution for the program:

1. The contribution the program will make within the current reporting period. This will not always exist—the program may not make any contribution until it is further along, or indeed until it is complete, but in some scenarios the interim deliverables (the completion of early phase projects) may lead to some incremental benefits. If these exist, then they will be a subset of the goals and objectives of the portfolio.

2. The interim contribution the program will make in future reporting periods. Again, this will not always exist and is essentially the same concept as #1 but for periods of time in the future. For example, if we are in the first year of a five-year program, then this category will cover years two through four—future periods, but not the final contribution. This is an area that is often ignored with the focus on the current portfolio and the final deliverables, but it needs to be considered by the program manager.

3. The final contribution. This is the overall benefit that the program is expected to deliver once complete and is the major reason why the program is being undertaken. This should be seen as the most important aspect of the program that program risk management is looking to protect (but not the only aspect).

We can try to use mathematical calculations (net present value) to compare quantitative values now and in the future, but as with the portfolio level risk management, we need to be able to apply qualitative analysis as well—the compromise between current and future objectives cannot be determined simply on objective values. Additionally, as we will see later in this chapter, the portfolio has significant impact on the program level risk management approach, and that can't always be predicted, let alone quantified.

There can be a tendency at the program level to compromise the second point above—the future interim deliverables. This is the result of either a reluctance to impact the immediate expectations when problems occur (at the program or portfolio level) or an equal reluctance to take any steps that will jeopardize the final deliverables. Instead, the *easy* option is to make changes to the interim periods. Examples may include moving resources off of the initiation or planning stages of those projects, moving budgets from those initiatives to current ones or deferring objectives from the current projects to later ones within the program.

This last example shows the dangers of this approach—the program only appears to be on track because the goal posts got moved to create that impression. In reality, most times this merely defers the problems and potentially makes the impact more severe; effective program risk management should be wary of any risk management activities that appear to be creating or supporting this situation.

When we consider the scope of program risk management, we also have to consider the potential overlap with project level risk management. There is a much closer working relationship between program management and project management, so there is a greater likelihood of confusion between the two than there is with the portfolio. It's important to avoid this confusion because if we don't, then we can either duplicate efforts in managing the same risk twice, or end up with risks that aren't being managed at all.

Generally speaking, a program will be envisioned before the constituent projects—a major initiative will be approved that will span multiple elements of the business and an extended period of time, and that initiative will be broken down into a number of projects with their own specific deliverables. This will mean that if the program is being proactively managed, there will be some risk identification, analysis, and perhaps even management activities underway before the first projects are initiated. As projects are scheduled and initiated, they will begin their own risk management activities, and this is where the potential for confusion and scope overlap can really begin. That's what we'll look at next.

Program Risk Management Downloading

If, when we initiate a project within a program, the project level risk management is conducted in isolation to the program level risk management that has already taken place, then there will be almost certain duplication and confusion. We therefore need to link these two processes together, and Figure 14.2 shows how this occurs.

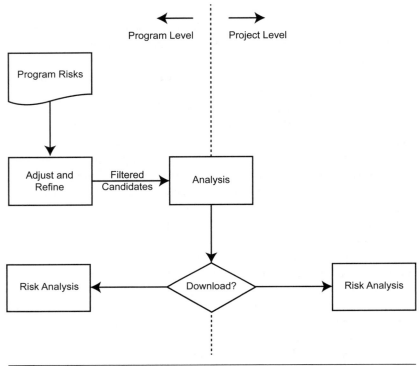

Figure 14.2   Program level to project level risk management downloading

As we discussed in Chapter 12, the adjust and refine process is where we interact with other organizational processes, so this is the logical place for stepping out of the organizational risk management cycle. We start by reviewing the risks that are associated with the program we are managing. During the adjust and refine phase, we are simply concerned with answering the question *is it possible that this risk is more appropriately managed by the new project that is being initiated*?

If the answer to that question is yes, then the risk becomes a candidate for *downloading*—moving the risk out of organizational risk management and into the risk management processes of the new project. Once we have completed that process for all of the program level risks, then we should have a filtered list that can be analyzed more thoroughly.

As you can see from Figure 14.2, the analysis process straddles the boundary between program level risk management and project level risk management; this reflects the fact that the analysis process needs to be conducted jointly between program and project management resources. This needs to be an expert review by equals with decisions made on merit, not positional authority—the wrong decisions will negatively impact both the program and the project.

The analysis process is concerned with establishing where the risk is most appropriately managed, and this requires a slightly different approach than with portfolio level risks. Because there is a lot of separation between projects and the portfolio, it's not generally appropriate to have portfolio level risks controlled at the project level. As we saw in the previous chapters, there may be a project level owner. However, that person works not as a stand-alone owner but with the portfolio level to determine the best approach and manage the risk and its impact.

The relationship between projects and the program that they belong to is much closer—there is less hierarchical separation, and there is a natural grouping around the program itself in terms of functionality and objectives. This means that it may be appropriate to have risks that are still completely program impacting fully managed at the project level. The analysis will investigate the following:

- Type of management. If the risk is being managed actively, then it may make sense to maintain a consistent owner rather than go through the transition of ownership at a time when a lot of management is occurring. This needs to be tempered to some degree by the consideration that the risk management activities may need to change now that the project is being launched.
- Most appropriate risk manager. Who is going to be the best person to manage the risk going forward. The initiation of a project will result in additional resources being engaged and a whole series of new project execution activities being undertaken, which may change the management approach and/or the ideal owner.
- Scope of the risk. If the risk is likely to remain beyond the completion of the project, then it should generally be retained at the program level. If it is downloaded to the project, then it will need to be uploaded back to the program when the project is completed, and that creates additional risk and inefficiency through two transitions. In this scenario, a better solution is likely to have a project level resource work with the program level risk owner.
- Alignment with project level risks. If there are similar risks at the project level, or if management activities for project level risks will deal with similar elements of the project, then we should try and leverage that and avoid duplication of effort.

- Status of the risk. If the risk is well under control, then there may not be any need to change ownership; that may simply cause additional stress to the process (a simple ownership change will increase the risk exposure to some degree while the transition occurs). At the opposite end of the spectrum, if risk management is having no impact, then this may be the time to change owners and perspective, even if it isn't the most natural fit based on other factors (although it may also be time to determine that the risk needs to be accepted and that management is simply not working).
- Project and program level capacity for risk management. This shouldn't be a determining factor (resources can always be realigned), but if there is no clear decision based on other factors, then this may become the final arbiter. If project level risk management capacity becomes an issue (and remember that there will be existing project level risks to manage as well), then we may need to undertake a prioritization exercise similar to that described in the risk analysis process in Chapter 9.

In all cases, we should be looking for clear evidence that downloading the risk to the project level is going to drive benefits for the management of that individual risk; otherwise the management should be retained at the program level. There is no point in moving the risk just for the sake of it.

Once we have determined whether the risk is best managed at the program or project level, the risk can return to the appropriate process—either the cyclical organizational risk management process or the more traditional linear process associated with project level risk management. In both situations, the risk should enter the process at the risk assessment stage. The initiation of a project that has some association with the risk is a fundamental change in the risk environment, and we need to review the risk to ensure that there have been no changes to the potential impact and that our risk management approach is still appropriate.

Risk downloading is an important step in program level risk management, but it is frequently avoided because the risk has been well managed at the program level; and there is a fear that the risk will not be managed as well if accountability is moved. Consider the following scenario as an example.

An organization has a program that is responsible for the major overhaul of their Enterprise Resource Planning (ERP) system. The program will involve a series of upgrades to different modules over three years and will include an infrastructure upgrade partway through the initiative.

The projects in the program are structured to address one of the modules each, and this reflects the different user groups, functionality, etc., for each module. There is a separate project planned for the infrastructure upgrade.

One of the identified risks is that the combination of new, more powerful functionality and additional users will put significant strain on the existing operational infrastructure. As a result, we have appointed a program level resource to work with technology operations to manage a number of key system metrics. If those metrics start to fall outside of acceptable parameters, then we may need to either accelerate the infrastructure upgrade project within the program, invest in hardware upgrades to the current system, or both.

Things have gone reasonably well, and the risk hasn't triggered; it is now time to initiate the infrastructure upgrade project that was planned within the program. Once this project is complete, the risk will be completely eliminated, but what should happen to the risk? Does it get downloaded to the infrastructure project or does it get retained at the program level?

The program level risk owner may well argue to keep the risk because until the infrastructure project delivers its functionality, the risk still exists, and decisions may need to be made quickly to avert problems. The risk owner might point out that the project is concerned with a new infrastructure while he or she is focusing on the impact on the existing infrastructure and is therefore the better owner with more focus on the current system. In many organizations, the program will retain ownership of the risk.

However, let's look at the criteria that we identified above for determining whether the risk should be downloaded.

- Is the risk being actively managed? Not really, it's just a monitoring activity that is being conducted so that shouldn't preclude downloading.
- Who is the most appropriate manager going forward? Well, while the program level resource seems to be doing a good job, the risk doesn't require a program specific skill set. Additionally the project has a lot of people engaged in it who are experts in technology infrastructure and the management of performance, so that seems to be a vote for project level management.
- Is the scope of the risk going to extend beyond the project? Clearly not, the project is specifically designed to eliminate this risk.

- Does the risk align with other project level risks? Obviously there is alignment, and further, the management of the risk may help to define the requirements of the project—providing an understanding of the current performance statistics and helping to define the extent of the required upgrades. These should come out of the requirements definition on the project anyway, but the alignment is clear.
- Is the risk already well under control? It appears to be, so this may argue against moving it; although as there is no specialist knowledge required to manage it, the risk of downloading is minimal.

We don't know the risk management capacity, but it seems clear from the above that the risk should be downloaded to the project. If that level of analysis is carried out, then it's generally simple to identify those risks that should be moved to the project level; the key is to ensure that the analysis is carried out rather than automatically retaining the control at the program level.

Program Risk Management Uploading

We've looked in quite a lot of detail at the idea of downloading—moving risks from the program to a project within the program under certain circumstances. We also need to consider the concept of *uploading*. Logically, this is the movement of risks from the project level to the program level, and it usually occurs in one of three scenarios:

- Management of a risk at the project level is no longer appropriate—this may be because the project level risk has been successfully managed, but program level exposure remains. It could be that project level risk management is not working, and a different approach is needed, or it may be that further analysis of the risk after downloading has caused a re-assessment. There still needs to be a program element to the risk in this situation—we aren't going to upload risks that are completely contained within the project.
- The project is being closed—if the project is wrapping up but there is still residual risk exposure, then the management of the risk needs to return to the program for management.
- The program is formed after the projects—if projects were initially approved as stand-alone initiatives and the decision to

consolidate into a program was taken later, then there will be no program level management infrastructure (for risks or any other program element), and these need to be uploaded from the various initiatives that form the newly created program.

If this last scenario is faced, then the uploading process is effectively the downloading process from Figure 14.2 in reverse—potential candidates for uploading come from the project and are analyzed using the same criteria detailed in the previous section. Based on this analysis, the decision is taken to either retain them at the project level or upload to the program, and regardless of which approach is taken, a re-analysis should be conducted to capture any changes in risk exposure, appropriate management approaches, and so on.

In the first two scenarios, risks will move individually or in small groups, and the process is simpler because the decision to upload the risk has effectively already been taken. These situations can be considered as little more than a transfer of ownership from an owner at the project level to an owner at the program level, although there will likely still need to be a re-assessment of the risk to identify any necessary changes in the management approach, or changes in exposure caused by the completion of a project.

We discussed in the downloading section that one of the elements that needs to be considered prior to committing to a download is whether the risk is likely to survive beyond the scope of the project, and if it is, then we should consider retaining the risk at the program level. However, we need to be realistic and recognize that at times program level risks will slip through and be mistakenly downloaded to the project, and those need to be pulled back to the program level. We also need to recognize that risks are fluid, and the situation may have changed significantly since any earlier download. Finally, there may be a risk that was never identified at the program level, originating instead within a project but with potential impact beyond that initiative.

The scenario that tends to be the most difficult for project managers to accept is when project level risk management is unsuccessful on a risk with program level impact. In many ways this can be compared to the overly optimistic project status report that many readers will be familiar with (a reluctance to report bad news) because there is a hope, no matter how unrealistic, that the project manager can recover the situation.

Of course, in project status reporting, that often results in the true status only being reported when the situation is so dire that recovery is nearly impossible. In risk management, the situation can be similar—the real situation with a risk being hidden through a misguided attempt to recover or to avoid admitting failure.

Program management needs to remain aware of all risks that are being managed at the project level, especially those with potential program level impact, and this can't be simply reviewing status reports. There should be regular risk review meetings that take place between the program manager and the project managers of all current projects within the program to review the individual risks and the overall program level exposure. If implemented and executed correctly, these should help to identify the early warnings of a risk that may need to be uploaded back to the program for management.

Resourcing Program Risk Management

In the previous chapter, we looked at resourcing portfolio risk management in quite a lot of detail. Much of the issues addressed there may also apply to program management, especially the concept of a team that is not dedicated to the program but is balancing multiple, potentially conflicting priorities. Just like with the portfolio, PMO resources may perform some of the program management functions—this commonly happens with some of the specialized functions that are not required full time on a program and can be shared from a central function like the PMO. From a risk perspective, risk analysis is an example of a function that may well fit this model. However, there is also a resourcing consideration unique to the program.

In some situations, the program management team may overlap with the project management function on one or more of the contributing projects. Some organizations may appoint the most senior project manager as the program manager, while others may intend to have a dedicated program manager but end up with that individual taking responsibility for a project based on resource availability, or as an attempt to help recover a challenged initiative.

While not ideal, this model can work as long as the individual is able to separate program and project level functions, and this can be difficult when the needs of program and project are potentially at odds. The program manager needs to challenge him or herself on all decisions to ensure that they are the right decisions for the entire program. This can be particularly difficult where risk is concerned if the decision that is right for the program may drive additional risk exposure, reduced risk management, or similar into the managed project.

There will also be scenarios where a program manager is responsible for more than one program at any given time, and this too can be a difficult situation to manage. The needs of both programs should be balanced, and this requires careful management of time and priorities. Of course in both of these situations, the program manager should feel that they can call on the portfolio management function and/or the PMO to assist with resolving conflicts and providing guidance.

Program Risk Changes and the Impact of the Portfolio

So far in this chapter we have been looking at the relationship between the program and the projects that make up that program, and this is where much of the day-to-day risk management activities will take place. The program will be considered as a self-contained unit that will be free to conduct all program execution processes within the process framework that the organization has defined. For risk management in particular this will be the application of the organizational risk management process elements and the subsequent adjustments that are necessary based on the changing project situations and the progress being made on the program.

However, as we know, the program doesn't exist in isolation. It is part of the portfolio, and at any given point in time change may be driven into the program level as portfolio management seeks to ensure that the overall portfolio goals and objectives are met. This may be direct risk impact—instructions to change the way that program level risks are being managed or new information that impacts the way that risks are handled. It is much more likely to be indirect impact—portfolio driven changes to other elements of the program (schedule, budget, resources) require a re-assessment of risk.

An example may be the decision at the portfolio level to move a key resource from a project within the program to another project that is not part of the program. This decision will have been taken with the intent of maximizing the likelihood of the portfolio achieving its goals, and there will have been some preliminary analysis done of the impact that will occur within the program. In many cases, there may have been discussions between the portfolio manager and the program manager before the decision was taken to assess the likely impact. However, when a change is actually executed, the impact still has to be assessed and the program adjusted.

In the ideal situation, the possibility of a portfolio driven change would already have been recognized as a possibility, identified, and assessed as a risk. In that case, all that is needed is the implementation of the contingency plan. This isn't as farfetched as it sounds. If discussion has taken place between the portfolio manager and the program manager about the possibility of making a change, then that should have caused the risk to be created or an existing risk to be re-assessed in preparation for any change.

Of course there will be many scenarios where the risk has not previously been identified, and we now need to deal with the situation. This can almost be considered a risk that has triggered before being identified, and we need to immediately execute contingency. As such, the key input will be the organizational constraints hierarchy that will guide us in determining which of the constraints can be compromised to protect the others.

Regardless of the actions that we take to deal with the specific risk that has triggered, we need to review the remaining program level risks to see whether any of them need to be re-analyzed as a result of the change. It is likely that some of the risks may now have changed risk exposure or require a change in management approach. There is also the distinct possibility that the resource who has been lost to the program was the owner of some of the risks, in which case those risks need new owners quickly to ensure continuity of management.

In addition to the portfolio driven changes, we need to also consider changes that the program needs to drive. These are usually the result of a triggered risk or a major change in exposure that impacts the program, and in many ways it is similar to the changes that we discussed for the portfolio level in the last chapter. We can use the same analogy from that chapter of a shock absorber; the program can attempt to absorb some of the changes in risks and protect other program elements from feeling the impact. However, because the program is closer to the projects than the portfolio, there is a greater likelihood that the impact of a triggered risk or a major shift in exposure will be felt—there is less capacity to absorb the risk. This is heightened by the inevitability of dependencies between projects that will make the *shock absorption* function one of minimizing impact, rather than eliminating it entirely.

This requires close communication between the program manager and the project managers, who will likely become aware of a potential issue at the same time. If that collaboration isn't in place, then there is the likelihood that both roles will act, but in different ways, or that each will assume that the other will act.

The program manager needs to drive the consistency in how a situation is handled. If the program manager believes that a change in risk exposure is a temporary situation and that no further actions are necessary beyond monitoring, then in most cases the project managers should likely take the same approach. There may be situations where some precautionary steps are taken on projects that have the potential for the biggest impact, but a comprehensive re-assessment should not be necessary.

Of course, because the program has to operate within the confines of the portfolio, the actions open to the program manager to absorb the changes in the risk environment may be tempered by the needs of the portfolio—elements of the program may have to suffer for the *greater good* of the portfolio objectives.

The Impact of Time on Program Risk

As we discussed earlier, programs have the unique challenge of being spread out over an extended time period. Decisions made today concerning risks may have minimal if any impact on the projects that are currently underway but may have significant impact on projects that have not yet started. This presents two significant challenges to managing risk within a program:

1. The impact of a risk on a project that has not yet started is difficult to predict. Beyond the scope and objectives, there may be minimal planning that has been completed on a future project within the program, and that makes accurate assessment of the impact extremely difficult. In some cases, it may not even be known that there is any impact on a future project because there is simply not enough information available. This in itself should be captured as a risk.
2. Risk management decisions tend to be made with a focus on the current problems. Problems facing the program today are much more likely to get attention than problems in the future. Even though organizational risk management is a much more strategic function than project level risk management, there will inevitably be a tendency to make decisions that are best for the problems that are being experienced today, not necessarily for the ones that have the biggest potential to prevent the program from achieving its goals.

When these two challenges are combined, the real difficulty of program level risk management is clear—we have to make the best decisions for the overall success of the program, but we don't know enough about the details of future projects to be able to accurately determine what the best decision for the program is.

To try and manage this risk, we need to consider completing some preliminary planning on future initiatives to at least understand major impact areas. This might be things like major work packages or potential vendors that provide some rudimentary framework to future initiatives and assist in understanding potential risk assessment. A subject matter expert on the deliverables of the program should also be part of the risk assessment team to help to identify the potential for impact of decisions on future projects. However, there needs to be an understanding that early on in the program (when a greater percentage of the work is still largely unplanned) there is a higher likelihood that decisions may have unforeseen consequences on later projects within the program.

As a result, programs with longer overall time frames should have larger management reserves to reflect the increased uncertainty that inevitably comes with those timelines. As the program and the projects become better defined and understood, the contingency reserves will increase through more complete risk identification and analysis, and the management reserve can be reduced.

15

Impact of Organizational Risk Management on Projects

This book is not about project level risk management, but we cannot ignore projects as part of the overall project execution environment. They are where the vast majority of portfolio work is ultimately completed and where the impact of portfolio and program level risk management is ultimately felt.

In the previous chapters in this section, we discussed the organizational risk management process and looked at the unique aspects created by the application of that process at the portfolio and program levels. All of this work is concerned with ensuring that the risks that may jeopardize the success of the overall portfolio or the programs within it are under control and that the organization is prepared to deal with the consequences of any risk that does trigger.

Projects are the execution vehicles for much of the work involved with this, and we looked at this to some degree when we discussed risk ownership during the risk management step in the organizational risk process framework during Chapter 10. Now we are going to look at the impact on project risk management in more detail.

Project Risk Management Fundamentals

Let's begin with a quick review of the basics of project risk management. At the outset of the project, the project manager will work with his or her team to complete a risk identification exercise aimed at identifying all of the potential risks faced by the project. This will generate a number of risks that need to be analyzed and prioritized before being assigned to an owner for monitoring and execution of any active management activities. Contingency plans will be developed, and during the execution of the project, the risk will either trigger, in which case the contingency is implemented, or management will be successful in preventing the risk from occurring, in which case there is no further impact.

Obviously, this is similar to the organizational risk management process that we discussed in Chapters 7 to 12, other than the fact that it is generally linear rather than cyclical. Unless there is a major change to the project, then there is no need to repeat the risk process; the project remains in the management phase until the project is completed. Even if there is a major change at the project level that requires a further risk identification and analysis exercise, this is simply another linear execution of the process, rather than the start of a cyclical approach—once the process gets back to the management phase, it remains there.

Other aspects of risk management are similar between the project and organizational levels. There will be (or at least should be) contingency reserves and management reserves that are calculated in similar ways, as we should see when we look at the organizational risk profile, and are consumed by the triggering of either the identified risks or the unforeseen problems respectively. There may also be a project level constraints hierarchy defined by the project stakeholders that helps the project manager to determine the appropriate management steps to control the project.

The analysis of risks at the project level is likely to be less detailed than at the portfolio and program levels. Instead of the extensive use of independent experts from anywhere within the organization (or beyond), there will be a reliance on the expertise of team members in their individual areas combined with historic project information. This is acceptable given the relatively low impact of project level risks and the recurrence of similar risks on multiple projects. There may be some situations where a more sophisticated analysis is required, but within the relatively confined scope of a project, the cost and effort involved with this level of analysis is likely only justifiable in extreme circumstances.

Portfolio and Program Driven Change

In the last chapter, we looked at the concept of *downloading* of risk management responsibility from the program level to a project within the program. In a number of places so far we have discussed the need for project level resources to be involved in the risk management activities for organizational risks, as they are best able to monitor what is happening, measure, and observe the impact of management activities, and identify any early warning signs that the risk is about to trigger.

However, these directly risk-related impacts are only a small part of the way that portfolio and program level risk management can drive changes into the projects. The attempts to preserve the goals and objectives of the portfolio and/or programs may cause virtually any element of a project to be impacted:

- Resources may be added, removed, or swapped with other initiatives
- Scope may be increased or decreased
- Delivery dates may be put back or brought forward
- Budgets may be adjusted upward or downward
- Quality standards may be changed
- Sequencing of deliverables may be shifted to ensure that a specific interim milestone is met
- The project may be cancelled, put on hold, or merged with another project

These are clearly potentially dramatic and wide-ranging changes, but they are perfectly valid in order to ensure that the more significant organizational goals can be achieved. Projects are effectively the pieces on the chess board that the portfolio manager will move around in order to try and achieve the ultimate goals; some of the less significant pieces may be sacrificed in order to ensure that *victory* is ultimately secured, with the portfolio achieving its objectives.

In considering how a portfolio driven impact will be accommodated, the project level constraints hierarchy should be considered in determining a specific course of action, and if there is a choice of options, the one that impacts a constraint that is lower in the hierarchy should be selected. For example, suppose that a project level resource is moved off a project and assigned to another initiative that is considered more important to the overall success of the portfolio. The project may be able to absorb that change either by reducing scope or extending the schedule, and the major driver of that decision should be the project level constraints hierarchy. If schedule is higher than scope, then the scope is reduced to preserve the schedule and vice versa. Of course, the other constraints will also factor in to the decision, and there may ultimately be a combination of different adjustments made to deal with the portfolio driven changes.

Note that in the example above it is possible that the highest project level constraint is resources, but a team member can still be lost by a portfolio driven decision because that is based on the portfolio constraints hierarchy, which may differ from the project level. If there is significant misalignment between the constraints hierarchies at the portfolio, program, and project levels, then it may indicate that the project priorities are not accurately aligned with the organizational goals and objectives.

Portfolio or program driven change should be incorporated into the project just like any other change—via the established change control process. However, there is clearly no need for any form of approval process—the project level change control board cannot reject a change that is driven from higher up the project organization. The change control board and project manager should feel that they are able to discuss alternative approaches with portfolio and/or program management if they feel that the goals of the portfolio and program can be met with less disruption for the project, but they can't drive that decision. At all times this should remain a collaborative effort between the various levels rather than an approach being imposed—think back to the concept of the risk management partnership discussed in Chapter 6.

Because these are significant changes, they will prompt a need to re-assess at least some of the risks that exist within the project, and they may also create additional risks that require analysis and management. These are potentially significant changes that can have a profound impact on individual projects, and dramatic change will itself drive risk into the project. At an organizational level, this should have been considered as part of the macro level decision on how to ensure that the portfolio or program has the best chance to achieve its objectives. At the project level, this ceases to be an abstract concept and now affects real people with real feelings and concerns.

Portfolio and program management functions should expect to meet with project teams that are affected by these changes and help them to understand why the changes were necessary and how important their individual contributions are to the overall success of the program and portfolio. Not only is that good leadership, it's good risk management.

Portfolio and Program Generated Risk Management

The portfolio and programs may have other significant impacts on project level risk management. There may be situations where, as a preventive measure against potential risk exposure at a higher level, they drive risk management activity into the project.

For example, consider a situation where a number of projects within a portfolio are scheduled to use the same vendor. The vendor has a great reputation, the organization has used them successfully in the past, there is not an excessive reliance on that vendor, and there are no indications that there will be any issues this time. There is some risk here because work is being carried out that is beyond our control, and that should be noted as part of the organizational risk identification. However, it's likely that the risk will be given a low priority at the portfolio level and is not likely to be actively managed because there are no *red flags*. A project that is using that vendor will likely give the risk a higher priority because the impact on that project in the event of problems will be more severe, but with a successful track record, there will not be any extreme management measures planned. On the face of it, this looks like a normal situation with a vendor we have relied on in the past.

However, the portfolio manager may decide that he or she wants to take some proactive steps to validate the assessment that the risk is low for this particular vendor, perhaps because there is to be a heavy reliance on the vendor later in the portfolio cycle. This may lead the portfolio manager to ask the project that is scheduled to be working with the vendor first to adjust the way that the risk is being managed. This might simply be asking for more frequent monitoring of the vendor at the project level and reporting to the portfolio on the risk status, or they may go a step further and ask the project to manage the risk extremely aggressively in order to *set the tone* for the relationship over the coming initiatives. This is intended to send a message to the vendor that the organization has heightened expectations compared to previous initiatives and cause them to step up their performance.

This is a hybrid form of risk management—the portfolio has still classified the vendor risk as low, but it is using project level risk management as a tool in an attempt to proactively manage potential future risk. If successful, this will lead to the vendor improving their own performance, resulting in a reduction in the portfolio level risk exposure—effectively outsourcing the risk management to the vendor so that they incur the costs of management.

This may sound a little underhanded, and it does need to be applied carefully—we don't want to alienate the vendor by pushing too hard, and we have to ensure that the project level risk owner has the appropriate skills and uses the right approach, but the tactic is perfectly valid. From the project perspective, there is an impact on the way that one particular risk is managed and reported on, but this is clearly still a form of project level risk management. We are primarily concerned with the performance of the vendor in their work on the specific project, and only that project. The organization is simply extrapolating that information.

The situation can also occur with the program replacing the portfolio, although here we have to be more cautious to ensure that the situation is truly isolated to the project and not a program level risk that needs managing. The closer relationship and greater number of cross project dependencies may mean that what looks at first to be just proactive risk management at the project level is actually an incorrectly assessed or prioritized program level risk.

Project Generated Portfolio and Program Risk Exposure

A portfolio will be made up of many different projects, and each of those projects will be exposed to many different risks. It stands to reason that some of those risks will have an unexpected impact at the program or portfolio levels when they trigger. In reality, these are portfolio or program level risks that were never identified, but this is not always going to be because of a failure of process at the organizational risk management level; rather, it is a situation where the impact beyond the project just couldn't have been foreseen—unknown unknowns.

The biggest problem with this type of risk is that even after the risk has triggered, there may not be a recognition that the impact of the risk is reaching beyond the project. For example, suppose that an upgrade to a software application caused an existing piece of functionality within that software to fail. This was never seen as a possibility, and the project team is still trying to understand exactly what happened so they can address the problem. The portfolio manager may well be unaware of the specifics of the problem—they'll just see reports that identify a delay and a use of the management reserve for the project. Because the failed functionality was preexisting, not something that the project was deploying, there is no identified dependency between this project and other portfolio elements, and because the risk had not been anticipated, there was no analysis of any potential impact. As a result, it may take several days to realize that there are other projects that are nearly ready to deploy that are reliant on the broken functionality working. Suddenly the portfolio manager is faced with several projects that are going to miss their deadlines, resulting in lost revenue, higher costs, and unhappy customers.

These situations will occur multiple times during a program or portfolio, and there is little that can be done to prevent them no matter how comprehensive our risk analysis. From a risk management standpoint, this is one of the reasons why we have management reserves, but we can't simply allocate some of those reserves and move on with our work. We need to recover the situation as quickly as possible and minimize the damage that is done to the portfolio; this is where communication between the risk management team is vital.

We talked in Chapter 6 about a risk management partnership, and that team should always be monitoring risks and triggers to make sure that there are no unforeseen impacts. When the triggered risk has not been identified—one of the *unknown unknowns* that consumes management reserve—the team needs to review the situation carefully. These are the scenarios that no one has analyzed and for which no one has prepared a contingency plan. As a result, they have the potential to have the most dramatic impact, and at a minimum we need to ensure that the potential impact on all aspects of the portfolio is identified and corrective actions initiated.

A consolidated analysis of all unforeseen risk events should be circulated to project, program, and portfolio management resources on a regular basis—no less frequently than weekly. It needs to be reviewed carefully as part of portfolio and program management activities. Any events that are suspected of having impact beyond the project on which they occurred should be immediately investigated by the manager who thinks they might be affected, so any necessary steps can be taken to minimize and recover. It's better to spend some time investigating something that doesn't have impact than to assume that there is no impact and only discover you were wrong when the situation is irrecoverable.

I have frequently incorporated this analysis with the adjust and refine process that we looked at in Chapter 12. Consider the review of triggered risks as a filtering process that identifies potential candidates for submission to the adjust and refine process for further work.

Web
Added
Value™

This book has free material available for download from the
Web Added Value™ resource center at *www.jrosspub.com*

16

The Role of the Project Management Office

There is one major group within project execution that we have not yet considered, and you could argue that it is by far the most important group—the Project Management Office or PMO. It's impossible to write at any length about PMOs without getting into a discussion of their role and structure because there are so many different models and spans of responsibility in use. We will touch on that but only to define what we will consider to be PMO functions for the purposes of this book. That may not match the functions that form part of your PMO, and your portfolio management and PMO functions may be organizationally part of the same group—that's okay. It's the specific responsibilities focused on in this chapter, not simply the job titles and reporting lines.

One final introductory point: while I'll continue to focus on risk management in this chapter, the concepts outlined here apply to the portfolio management methodology as a whole and not just the risk management process. It's impossible to talk about organizational risk management without considering the related processes that make up the overall portfolio execution methodology, so you will find more frequent mention of portfolio management as a whole in this chapter.

A Note about EPMOs vs. Traditional PMOs

In recent years, the concept of the Enterprise Project Management Office, or EPMO, has been gaining a lot of traction. At the risk of dramatically oversimplifying them, they are a consolidated, enterprise-wide PMO that seeks to leverage economies of scale, standardize project execution approaches across the entire organization, and allow for improved effectiveness in project execution. They are frequently formed by consolidating department or division-specific PMOs into a single entity.

EPMOs are a natural fit with the concept of a portfolio execution approach—both are organization wide in scope, and the standard approach is exactly what is needed within the portfolio. That doesn't mean that portfolio execution can't be successful within a distributed PMO model. The project level processes can be different as long as the organizational processes are the same. In fact, even under an EPMO there will be variations in the project level execution approach; for example, individual initiatives may follow Agile processes where appropriate.

If the organization has an EPMO, even if that is currently in addition to department level PMOs, then ownership of the process elements that we are going to discuss here is clear. If the organization still has a number of self-managed PMOs operating at the same level within the organization and contributing to a single organizational portfolio, then there needs to be an owner established for those processes—a single PMO that will act as process owner for organizational risk management and the associated organizational/portfolio level processes. This does not put that PMO higher in the organizational hierarchy; rather it gives them an additional process ownership responsibility. While the decision on which PMO owns these processes can be somewhat arbitrary, it should be a group that has the experience with projects and a fairly broad presence within the organization.

If on the other hand the organization is structured with department level PMOs each running their own portfolios, the situation may be more complex. There is absolutely no reason why the concepts discussed here can't be successful in that environment, but before an organization deploys a portfolio risk management model at the department level, it needs to be sure that it intends to keep that segregation of portfolios. Trying to consolidate department level portfolio execution once each department has a different portfolio execution methodology in place can be incredibly challenging. A portfolio should be an organization-wide entity, but I recognize that is not always the reality.

PMO Functions Supporting Risk Management

In looking at the role of the PMO, I want to focus on a number of different functions that will assist with organizational risk management. Some of these have been briefly discussed within the previous few chapters, and some of them are new here. The PMO functions are varied, but they share the common thread of being supporting—they form the framework in which all portfolio management functions, and for the purposes of this book, specifically organizational risk management, can operate successfully. I am going to define a few parameters for these support functions:

- They are all applied across all levels of the project execution organization—portfolio, program, and project. The specific elements may differ at the various levels, but there is a role for each function to play at each level.
- They are indirectly associated with the organizational risk management process. While all of these functions are required to maximize the chances for success, they are not directly involved in the execution of the processes themselves—those have already been addressed in the previous few chapters.
- They are independent. The portfolio and its constituent parts will have many different stakeholders, all of whom will have their own personal agendas for what they want to get out of the portfolio. The PMO functions are focused solely on maximizing the chances of success for the portfolio in terms of the defined goals and objectives and will never *take sides*.
- They are advisory, not prescriptive. The role of the PMO support functions is to strengthen the overall risk management (and overall portfolio execution) framework, but the ultimate decision maker will be the portfolio manager. PMO functions will advise and make recommendations but will ultimately comply with the portfolio manager's decisions.

Each of the steps above adds to the chances of success for organizational risk management, but that's not the same as saying that risk management cannot be successful without some of them being in place; it will just be harder to accomplish because the framework for success will not be so well defined.

Similarly, if you have these functions in place, but they are located other than in the PMO, that's okay. The PMO has accountability for them, but it's not necessary for that to translate into direct control.

Process Ownership

This isn't so much a function that the PMO performs as it is an ownership role it has. The PMO should own all project related processes, whether at the portfolio, program, or project levels. Process ownership means that the PMO has:

- control over the project execution methodology, all of the processes that contribute to that methodology, including templates, process flow diagrams, etc.
- ownership of process review and continuous improvement work, including decision-making authority on changes to processes
- responsibility for training users on how to use the processes
- control of process audit functions to monitor compliance and address variances.

The PMO does not own the application of the processes within individual situations—that is the responsibility of portfolio, program, or project managers.

The independence of process ownership from process execution ownership is important as it gives an unbiased view on the functions that simply can't be achieved from someone who is embedded in applying the processes, and the PMO is the logical place for this ownership to reside. The only reasonable exceptions that I have seen to PMO ownership is in organizations where there is a distinct process development team that owns all organizational processes. In that situation, the PMO will act as a subject matter expert that works with the process team.

Organizational Culture

In the final section of this book, we are going to talk a lot about culture as we look at how to successfully implement an organizational risk management approach within your organizations, but let's set the groundwork here. Fundamentally, no process, policy, rule, or guideline will be successful within an organization unless there is demonstrated consistent commitment to it. That can take a long time to establish, and it can be destroyed in a moment if not truly ingrained into the way that an organization operates—it needs to become a part of the culture, part of what shapes the organization.

This may sound melodramatic, but it really does need to be that integral to the operations of the business—cultural change doesn't occur when the project manager says no to a sponsor who asks them to bypass process in an attempt to save time. It happens when the sponsor knows that it isn't an appropriate question to ask.

Cultural change cannot be imposed; it has to be accepted by the organization, so the role of the PMO in its capacity as the process owner is to create an environment where the culture can evolve and embrace organizational risk management and the larger portfolio driven project execution model. It requires the following tenets:

- Consistent application of process—making sure that all of the process steps are applied in full, that variances are addressed and corrected, and that exceptions are only allowed when fully documented and approved
- Open and visible information—while a small percentage of an organization's initiatives will always have to be conducted in secret—due diligence for a potential corporate takeover for example—the general approach should be to have as much transparency as possible into the results of process audits, outputs of process, and so on
- Equality and collaboration—treating all stakeholders in the process as equals and embracing changes that will help to evolve and improve the execution approach, regardless of the source
- Contextual process—ensuring that all stakeholders are aware not just of what processes are, but why the processes exist and how they support the organizational goals

If these tenets are applied consistently, then over time the organization will begin to embrace them as part of what makes that organization unique, and the processes will become part of the *organizational DNA*.

Organizational change cannot be rushed—it takes time to be understood, longer still to be accepted, and even longer to become an integral part of the organization. If the processes that make up that change are low impact and/or frequently executed, then the process of acceptance can be fairly quick—it doesn't take us long to adjust when our password expires and we have to select a new one, for example.

However, when we are talking about major elements of the way that projects are executed, the changes are far more significant and less frequently executed. It can take years for acceptance to occur, and throughout that period, the PMO has to ensure that the process is being consistently applied, addressing any problems as soon as they arise. If one initiative gets away with not applying risk management process, or if one set of risks are rushed through without proper analysis, news of that will spread across the organization immediately. The next time that the PMO tries to enforce process they will be faced with questions along the lines of "how come the risk management process was skipped on x project, no one signed off on that as an exception and they got away with it?" Apart from the fact that no one wants to have to answer that question, it establishes a precedent that undermines the chances for successfully changing the culture.

If there are a lot of closed door meetings going on within an organization, there will be a lot of rumors circulating among employees not involved in the meetings. There will be a lot of guessing about what is being discussed, and all of it will assume that something bad is going on. The same will happen if there is a lack of visibility around the application of process for a particular part of the portfolio—the assumption will be that the processes aren't being followed, and again the embedding of the approach into the organization is undermined.

Cultural change can never occur without the people involved feeling a sense of ownership—they need to feel that they are an integral part of the process, that they have the ability to help shape and improve it, and that their voice is just as important as anyone else's. If this is achieved (we'll look at some ways for the PMO to do that later in this chapter), it helps create the cultural change. If employees feel that the process is being *done to them*, then the opposite will occur, and people will start to reject the changes. Employees also expect that they not only have a voice, but that it is heard—a PMO that encourages submissions for improving process and then rejects every one of them will get a reputation for only paying lip service to wanting to include stakeholders in developing the process.

The final piece of this is the importance of providing a context for the process within the organization. True ownership, along with the empowerment and motivation to drive improvements, will only come with understanding of why processes are being applied. When we manage projects, we try to ensure that our team members understand the reasons behind the project as a whole and their part of it in particular. That way they feel as though they are working on something meaningful; they understand how their work contributes to the project as a whole.

The same happens here. If people understand how the hours that they spend analyzing risks and monitoring and reporting on a risk that never seems to change contributes to the overall success of the portfolio, then they are more willing to accept and embrace the work. Asking someone to check on the number of system errors every day may result in compliance but is hardly a motivating task. Asking that same person to monitor the number of system errors every day because if the average in a week rises above 10 per day, it could be a sign that the system is failing, which in turn could delay the company's most important product and result in the CEO having to explain to the board of directors and shareholders that they won't make their targets this year, will have a much more profound effect. Not only will the employee make sure that they are checking the system, they will feel as though they have a vital role to play in the portfolio's success.

Even with all of these measures in place, there is no guarantee of cultural change occurring. I witnessed a situation where two vice presidents in a company were walking down a hallway. One said to the other, "This portfolio management thing is a joke. There's no way that I am going to give up control of my projects to someone who doesn't understand my business." The outcome was predictable—the portfolio manager was undermined, the processes were seen to be a waste of time by all levels of the project execution framework, and the PMO was in damage limitation mode trying to reestablish the benefits and demonstrate that an integrated, process driven portfolio management approach really could work.

This sounds like a whole lot of doom and despair—success needs cultural change. That takes years and can be destroyed in an instant. There's a lot of good news as well—every organizational risk that is successfully managed contributes to improvements in the organization and a greater likelihood of portfolio success. However, that shouldn't be accepted as being good enough. Organizational risk management should be seen as just that—the standard for the entire organization and every single risk. That inevitably will take time, energy, and commitment to achieve, and continuous active management to maintain.

Education and Training

There is a lot more to successful project execution than simply following each step in a process. No matter how well written the process is, or how clear the templates are, the mechanical execution of steps without any thought or interpretation will never yield successful outcomes—the real world is simply not that black and white. A number of elements of the PMO's role impact on this, but the biggest is the provision of education and training. There are two distinct elements to that:

- The skills, knowledge, and judgment to work successfully within the organization—training on what is expected from project (or program, or portfolio) resources within the unique organizational environment.

- The detailed process training that provides the context for the application of the skills developed in the point above.

Let's consider each of these separately.

Skills, Knowledge, and Judgment Training

The PMO is responsible for ensuring that all of the people involved in the execution of projects have the skills and knowledge required to be successful, together with the ability to apply those skills appropriately. This effectively means that the PMO ensures that each individual within the project execution organization—from the portfolio manager to the most junior team member—has the skills to be able to do their job properly.

Clearly the specifics of the training will vary from individual to individual but will consist of some combination of the following:

- Project management training—the provision and enhancement of project management skills to allow an employee to perform the tasks expected of them. For team members, this may be internally delivered through fairly informal training classes focused on terminology, how to estimate, and types of dependency. For project and program managers, it might be more structured; externally delivered training courses teaching specialist skills like quality, scheduling, cost management, and risk. This may also include the requirement for a formal certification like the Project Management Institute's Project Management Professional (PMP) or Program Management Professional (PgMP) designations.
- Leadership and soft skills training—similarly structured to the project management training, this will focus on the ability for managers and senior team members to be able to communicate effectively, coach, motivate, and generally lead their team members in a way that maximizes the chances for success and facilitates high-performing teams. There may be some basic communication and confidence type training provided to all team members—the most junior team member may end up owning a risk that has the visibility of the portfolio manager.
- Organizational approach training—regardless of the experience of the individual, or their proven capabilities in the previous two categories, when a new resource joins an organization, they need to understand the way that projects are executed within the organization. This is foundation training for the

process and methodology training that we will look at next. It concentrates on the high-level concepts of how projects are considered and approved; how the portfolio is structured and executed; how project managers and team members are assigned; how different PMOs work together (if there isn't a central EPMO); and high-level concepts around governance, estimation, planning, change control, and all of the other project disciplines. By providing this basic framework to everyone, there is a high-level understanding of how all of the different project execution elements come together. While not everyone will need the detailed process training on every element of the processes, there will at least be a basic understanding of the overall methodology.

Much of this training will be provided when a new hire joins the team or someone transfers in from another part of the organization, although skills-based training will be an ongoing program as people gain experience and responsibilities.

A lot of organizations make the mistake of choosing training vendors, or developing their own training, without consideration of the ability to apply the skills that are learned—the conceptual ability is not sufficient, it's the decision making around how and when to apply those skills that makes the difference. To use a risk scenario as an example, it's not sufficient to know that the four strategies for managing negative risks are acceptance, mitigation, transfer, and elimination—you have to be able to determine which one is appropriate in each situation. Training needs to include exercises to practice this judgment, and the timing of the training needs to allow the resource to be able to apply the skills within a reasonable time frame. Sending someone on a three-day training course to learn about risk and then not assigning them to any risk management activities for six months is not going to help!

Process Training

Process training is specific to the organization and the way that projects are executed—it assumes that people already have the fundamental skills from the previous section. To use organizational risk management as an example, training would focus on these items:

- How the organization views risks—the high-level risk categories that apply to the specific organization within its industry, geography, and regulatory framework.

- The idea of the risk profile and how the organization monitors and reviews organizational risks
- The framework of organizational risk management—the organizational constraints hierarchy, the portfolio's focus on managing risks against the ability to achieve the organizational goals and objectives, and the concept of a risk management partnership across the portfolio
- The organizational risk management processes themselves—the concept of the cyclical approach with the ability to break out into related process areas, along with each of the individual process elements

Once these elements of process training are complete, there should be a good understanding of how the process should be executed, and this is the point where many PMOs will stop training. Often this type of process training is based on the process documentation itself and may consist of nothing more than reviewing process flows and templates and allowing trainees to practice on a few risks.

If we stop at this point, we will end up with a group of resources who are capable of applying the developed processes effectively, and for many organizations that is all that they want to provide. However, to maximize the chances for success, we need to go further and learn from the inclusion of judgment training in the section above. Once we have trained people how the process works, we have to train them to know when to bypass the process.

That may sound odd given how far we are into this book and how much time I have spent detailing the various processes, but we have to recognize that we operate in an imperfect world. Processes may provide us with a framework that works 99.9% of the time—a stable, repeatable process that is consistently applied in order to maximize our chances of success. However, there is that 0.1%—that exception scenario where we do have to *break the rules*. The key is to ensure that we can recognize when that 0.1% situation exists and that we break the rules in the right way—and that's where the training comes in.

By definition the situations where we can bypass process are exceptional, so training needs to focus not on specifics but rather on the type of scenario where ignoring process would be acceptable. Typically, this would be an extreme scenario where timeliness is crucial. For example, if a bug was found in a piece of software that jeopardized the integrity of users' personal information, then we may bypass all of our normal maintenance processes and just turn the system off without any notice to users.

From a risk management standpoint, it is possible that a situation like this will occur when a risk is triggering, and we need to act quickly to prevent a severe impact. This is already anticipated to some degree with the development of contingency plans.

Process training will focus on helping risk management resources understand how exceptional this situation is—it's not going to happen once a month, once a project, or even once a program, but at any given point it could happen, and training helps people to recognize and respond.

The second part of training for exceptions is to ensure that there is training on how to capture and document the fact that the process has been bypassed. Clearly if time is critical, we don't want people to stop what they are doing and fill out a form explaining why they are not following the proper processes, but there does need to be an awareness of an exception process. Typically, this will be a standard process that will be followed for any process exception, whether it is related to risk, quality, schedule, or any other process area. It will generally consist of a simple explanation of what happened and why process was bypassed that is circulated to stakeholders for review and sign-off and is then submitted to the PMO for archiving and review to see whether the situation raises any opportunities to improve processes. The more detailed explanation is still important, but that will be completed once the initial crisis has passed. Again that will follow a standard that is common across all process areas, not just risk management.

Process Audit and Control

As the owner of the project execution processes, it also follows that the PMO owns the audit and control function for those processes. The degree of formality of these audits will vary considerably from one organization to another, or even across different PMOs within the same organization, and the amount of control that the PMO can exert is equally varied. While both loose and tight audit and control functions can work equally well, there is a need for clear communication around that role to try and avoid resentment and conflict between the PMO and project execution teams. Auditing is never going to be welcomed by the group being audited, but if there is an understanding that the focus is on the process and not the individual, then there will be a greater acceptance.

Although auditing and control are related functions, they are focused on different aspects of process execution:

- Control is concerned with trying to ensure that the processes are executed properly. It is a predominantly proactive function that is aimed at the prevention of variances, or at recognizing and correcting variances early. Control can be considered as establishing the framework against which audit is measured.
- Audit is concerned with measuring compliance of execution with the processes. It is a predominantly reactive function that is aimed at the identification and correction of variances after they have occurred. Audit can be considered a measure of the success of the control functions.

Audit and control cannot exist in isolation. To be of benefit to the organization they have to be conducted in conjunction with a process improvement process that we will look at later in this chapter. Let's look at the control and audit elements with particular focus on the organizational risk management process.

Control

Because control is focused on prevention and early recognition, some control elements can be built into the process. When we detailed the organizational risk management process, there were a number of review and approval steps—at the completion of risk analysis, for example. These act as checkpoints to prevent the expenditure of time, effort, and money until we have confirmed that the approach we are taking is appropriate—they form approval gates within the process. For our purposes, they also serve to demonstrate the difference between the process and the work—between the PMO responsibility and the portfolio management responsibility.

An approval step in the process serves as:

- A *control* point in the *process*. It is the PMO's responsibility to ensure that the process has appropriate control points in place to validate the work that has been completed to that point and make corrections prior to a significant commitment of time, effort, and/or money.
- An *audit of the work performed* in the *execution of the process*. It is the portfolio or program manager's responsibility (in the case of organizational risk management) to ensure that the approval process is executed appropriately for their portfolio or program.

Note that while we refer to this second bullet as an audit, this should not be confused with the PMO audit that we will discuss next. Here, the audit is of the work performed—it is the execution of a process control function, not an audit of that process.

Review and approval steps are the most obvious examples of process controls, but they are expensive to execute, involving inevitable duplication of effort as people check the work of colleagues to validate the steps taken. As such, these controls should only be used in situations where there is the highest exposure to the organization if mistakes are made— the high-risk process steps in other words.

In other places within the process, other controls are appropriate— lower cost management for lower exposure risks, if you will. One of the common process control techniques used here is the checklist, often combined with a requirement for signature by the person performing the task.

Think back to the section on qualitative risk analysis in Chapter 7. We said that qualitative analysis had to consider:

- The likelihood of the risk occurring
- The impact that the risk will have on the portfolio and/or programs if it does occur
- The effect that management will have on the risk exposure (likelihood to occur and/or impact)
- The cost of management
- The ability of the organization to manage the risk

If our risk assessment template has an instruction or cover sheet that outlines each of these considerations, then every time someone assesses a risk, they will see this list of items to consider, effectively providing a control point by reminding the team member of the items to consider. If we wish to formalize the control more, we can provide a formal statement for the risk assessor to sign that specifically confirms that all of these items have been considered during the analysis. We have to be cautious with the use of a sign-off. If there are too many sign-offs required during the process, then they will lose their significance, and resources will just automatically sign them without considering the significance. However, used appropriately, they form an effective and inexpensive check on the execution of the process.

A less invasive, and potentially more successful, way of incorporating checklists into the process is to use a more intelligent template design. Consider the example that we just looked at above. Instead of having a checklist that the risk assessor has to review, the template that they complete could have a section in it for each of the considerations. This will automatically get the assessor thinking about each of those aspects of qualitative risk management and eliminates the possibility of something being missed.

The flip side of this is that if we break a process element down into too many arbitrary steps, then we may end up driving the wrong behavior. In this example, if qualitative analysis of a risk simply requires five sections of a template to be completed (along with some basic risk identification items), then the focus will quickly become completing those sections and moving on, not on ensuring that the underlying risk is subjectively analyzed and understood. We end up trading completeness and accuracy for convenience, and that in turn drives additional risk into the process execution—the exact opposite of what we are trying to achieve.

Controls often don't focus on the process elements themselves, but on the timeliness of completion. Most project processes are time critical. With risk management in particular, all of the time that a risk is less than fully understood and is not being managed, there is the potential for significant exposure with no warning that the problem is about to occur. Processes should therefore establish guidelines for the time needed to complete each step, and these guidelines will help the individuals executing the process to understand how much time they should spend on each step. The control should include a recommended minimum and maximum time—we don't want an item to get stalled in the process, but neither do we want the process step to be rushed. To consider our qualitative analysis example, it is quite easy for *analysis paralysis* to set in as we try to better understand the nature of the risk. However, if there are a large number of risks requiring assessment—straight after an annual planning cycle, for example—it can be tempting to rush the analysis in an attempt to bring an individual workload down to a more manageable level. Care needs to be taken with the use of time-based controls, the time ranges should represent best case and worst case scenarios. However, often they turn into norms, with every instance of the process execution either being rushed or prolonged. The addition of an average time for completion can help that situation by establishing a norm that is in the midpoint of the range.

With the increasing use of project portfolio management (PPM) software, or of workflow modules within enterprise software solutions, process control has become both easier to implement and more flexible in application. As soon as templates are moved into an integrated software application, we can apply automated checks to ensure that every field has been completed or that the declarations have been signed off, for example. We can also apply automated warnings if process steps are taking longer than expected, or if a task appears to have been completed sooner than is considered normal. PMOs should look to leverage such tools when they are available to them, but they should never replace the good judgment needed to apply appropriate controls to the process.

Audit

As we saw before, the control mechanisms can have elements of audit in them, but there we are concerned with auditing the work performed. In the case of the PMO's audit function, we are concerned with a process audit, and that consists of a number of different elements:

- *Process compliance.* How well did the people executing the process steps comply with all of the requirements of the process, following the process steps, completing the appropriate templates, and meeting the expected timelines?
- *Process effectiveness.* How well did the process achieve the goals that it was designed to achieve—in the case of risk management how effective were we at identifying, analyzing, prioritizing, and managing the risks, as well as dealing with the impact of triggered risks?
- *Process efficiency.* Was the process efficient in terms of time, effort, and cost? Was the execution of the process incorporated into other portfolio execution work without significant impact? And did the cost/benefit analysis show an appropriate return?

These are unlikely to be undertaken as separate audits, rather there will be a single comprehensive audit that will consider each of these elements. There will also be two distinct comparisons for each of these elements:

1. The absolute performance—whether the measures are acceptable against the expected standard
2. The relative performance—how the measures compare relative to previous projects, programs, or portfolios

Before we look at the details of the audit, let's deal with one item right up front. Process audits should be about the success or otherwise of the process, not the people performing the process. If there are deficiencies, then they are likely going to be isolated to poor communications, poor training, or maybe just a plain bad process. The vast majority of employees are going to do their best to meet the expectations set for them, even if they don't always strive to exceed those expectations, and consistent failure is unlikely to be a problem with an individual or team. This needs to be understood by both the PMO and by the people executing the processes being audited.

The most common audit element is a compliance audit. This will generally look at the documentation that results from the execution of the process and identify any shortcomings in the way that the process is applied. Documentation is generally in the form of completed templates, and the process audit will be looking for certain items:

- Missing templates, indicating that a step in the process may have been bypassed
- Incomplete templates, indicating that a step in the process may have been only partially completed
- Inappropriately completed templates—wrong information, brief information, or excessive use of terms like *not applicable*, indicating that the process may either have been rushed or misunderstood

A process compliance audit should not focus on one or two instances of the process being executed—there are far too many variables for that to be meaningful from a process analysis standpoint. Instead, the focus should be on identifying trends—the one or two process elements that caused consistent difficulties, the one or two team members who seemed to struggle in multiple process areas, or the one or two template fields that were frequently misunderstood. This will provide an indication of a systemic problem that needs to be addressed.

More detailed root cause analysis will be needed to help establish the exact nature of the problem, and care should be taken in making assumptions based on the evidence. It's easy to assume that if one or two people are consistently having problems with compliance then they need more training on how to follow the process properly, and that may be the solution. However, the root cause may be that the training they received was inadequate, the guidance they are receiving from a manager is inaccurate, or they are relying on an out-of-date manual/process guide.

We also have to consider the relative compliance compared with previous audits. If we are seeing a repeat of the same problem areas, either we haven't identified the true root cause or the corrective measures we implemented to address the compliance shortfall were insufficient or incorrect.

The next audit element to be considered is a process effectiveness audit. This looks beyond the concept of whether the process is being followed and considers whether the execution of the processes has achieved the organizational goals. In the case of organizational risk management, we will be concerned with ensuring that the steps we took were successful in the following:

- Identifying the risks accurately, completely, and in a timely manner
- Analyzing the risks accurately
- Prioritizing the risks appropriately
- Identifying the right manager, management approach, and contingency
- Managing the risks effectively throughout the initiative
- Determining accurately and in a timely manner that the risk has triggered
- Implementing contingency, impact assessments, and recovery effectively

In an effectiveness audit, we are looking for areas where the process was unable to achieve its objectives, and we are trying to understand why those failures occurred. This clearly has to be considered in combination with the compliance element of an audit. If the step is not being followed correctly, then there is unlikely to be an effective outcome (although if there were consistent shortcuts taken in process execution without impacting effectiveness, then the process is likely over engineered).

When looking at process effectiveness, we have to look beyond the pure process steps themselves and consider all of the variables. Suppose that our risk analysis was consistently ineffective—failing to accurately identify the risk exposure, missing some impacts entirely, and mistakenly identifying exposures that didn't exist. To get to the root cause of why the analysis didn't work, we have to consider several things:

- The process itself—the details of how to conduct qualitative and quantitative analysis. These may have been inaccurate or unclear, which could result in ineffective execution.
- The information sources identified as inputs to risk analysis— things like the project archives, systems that provide data for quantitative analysis, and expert areas within the organization who can assist. If some information sources were left out, if some were wrong, or even if there were too many sources without any categorization to assist process practitioners, then execution could easily have been ineffective.
- The process support infrastructure—the guides, templates, training, and help available to team members to steer them through process execution. If these were incomplete, unclear,

missing entirely, or not available when and where they were needed, then the process is likely to be less than successful.
- The resource selection process—the way that the people executing the risk analysis were selected. If there are not clear guidelines on the skills needed by people who will be conducting analysis, or if those guidelines are not followed, then failure is likely—just like any other project task, success requires the right people to be assigned to the work.

We can see from this relatively simple example of just one step in the risk management process that effectiveness audits may drive a number of changes to the process and supporting elements. There may be a need for changes to the process itself, but in many cases, improved effectiveness relies on improving the way that the process is executed. This element of a risk audit requires more interpretation than the relatively black and white compliance element. Here we need to determine how best to improve the effectiveness—through the use of better controls, better support infrastructure, better process, or something else entirely.

This makes the use of relative effectiveness even more important. By comparing the results of one audit with a previous one, we can see whether the steps taken between the two were effective and help to determine how best to improve things this time around. We also need to recognize that sometimes the changes that we make can take some time to become embedded. If people have been used to doing things a certain way for some time, then it can take them a few iterations of the process to adjust to the new way of doing things.

We must also be careful not to implement change for the sake of change. Effectiveness should ultimately be measured against a standard—it is not realistic to expect that every time the risk analysis process is executed, it will result in a perfectly accurate analysis of the risk exposure that the organization faces. The PMO needs to establish what that standard should be and measure effectiveness against it. For example, if 95% of the risk analyses conducted for a particular portfolio were judged successful, then most organizations will deem that to be a successful process and will not look to make significant changes. It's always worth looking at the 5% that failed for any common issues to pick (and fix) the low-hanging fruit.

If we determine that we need to improve performance to 96% or higher, then there will likely be a need for significantly more rigorous process and much more detailed analysis, which of course will cost more time and effort, as well as requiring more time for the work to be conducted.

This becomes an exercise in risk management—do we want to commit the resources to risk management in order to mitigate the risk from a 5% failure rate to a 4% failure rate?

There's no simple answer to that. The organizational risk tolerance becomes a factor, as does the risk capacity, and on any given project or program we may need to adjust the acceptable standard. The PMO's role is to ensure that the organizational standard against which effectiveness is measured is appropriate for the organization's current risk profile.

The final element of a process audit is process efficiency, which considers the cost of executing the process. This looks at the financial cost, the effort cost, and the elapsed time for the execution of each process element to identify areas where the process can be made more efficient. We generally focus on averages here—the mean time taken to complete a risk analysis along with the average dollar and effort cost. This has become much easier to track in recent years with PPM and workflow tools that can automatically capture these data elements without the need for expensive manual tracking or inaccurate self-reporting.

There are many variables that can impact the efficiency of a process— complexity, experience of resources, and workloads for example, and when we seek to compare efficiency with previous audits, we also have to consider any changes in the processes and in the standards for effectiveness. As we saw, these can drive additional work, which will appear to hurt efficiency.

Many organizations will focus on relative efficiency, setting organizational goals for the PMO to improve efficiency by x% each year. I find this approach too simplistic and much prefer a specific standard for efficiency that is set in conjunction with the standard for effectiveness. This then allows for a reasoned decision-making process that consciously trades off effectiveness and efficiency (more effective generally means less efficient and vice versa) and still provides a solid target for the PMO. The danger with a relative improvement expectation each year is that we end up focusing on areas with diminishing marginal returns, instead of diverting resources and focus to areas where significant improvements can still be made.

Risk Audit

In addition to the process audit and control concepts that we just reviewed, the PMO has a different audit function to perform. This has nothing to do with the PMO's role as process owner; rather, it is focused on the PMO's responsibility to ensure that the projects being executed within the organization are managed appropriately. This will cover a lot of different areas—financial, schedule, quality—but we are going to focus on risk specifically, as that is our concern here.

The PMO's risk audit function is intended to be an independent review of the risk management that is in place in the portfolio, programs, and projects in order to ensure that there are no issues missed that could derail the initiatives. This is one of the most unpopular PMO functions because project execution team members frequently view it as second guessing of their work and consider it to be a lack of trust in their abilities.

In truth, this function is necessary to ensure that there is an objective view of the work that has been done. The people working on specific initiatives are deeply engaged in the work and will always look at that work from their own perspective as part of the team. This cannot be completely objective, so the PMO performing an independent audit provides the objective validation, along with any adjustments that are needed.

Risk audits will generally be high level; they are not concerned with validating every decision that has been made; rather, they are looking at the high-level data and seeking to confirm that it is broadly in line with the summary information reported. In the case of a portfolio level risk audit, the focus will be on these items:

- Assessing the overall risk exposure, contingency, and management reserves, along with the effort and money invested in risk management. This will be done through a random sampling of the assessed risks, the prioritizations given, management approaches, etc. The audit is concerned with ensuring the decisions are logical, the management is consistent with the exposure, the current status is being maintained and updated, and risks are being triggered when appropriate.

- Monitoring alignment with the organizational constraints hierarchy to ensure risk is considered at the appropriate level within the hierarchy. This will again be a random sample, looking to ensure that the approaches to management never require the compromise of a constraint that is higher than risk in the constraints hierarchy; and that where needed, compromises are being made in constraints that are lower in the hierarchy to minimize the impact on risk.

- Monitoring risk outputs to ensure that they are aligned with the process. This is done through a random check of documentation to ensure that contingency plans are in place, the risk management plan is current, and that individual risk summaries accurately reflect the current state. This item has similarities to a compliance audit but is conducted on a random basis throughout the portfolio execution and is intended to flag opportunities for immediate action rather than systemic process problems.

Throughout a risk audit, we are focused on the specific data for each risk—the process audit looks at the *how* of process execution and this risk audit looks at the *what* of the risks that go through that risk process. Just as with process audits we are focused on trends, but here we are looking for issues with the risk data that are resulting in unnecessarily high-risk exposure, ineffective risk management strategies, contingency plans that won't work, incorrect risk triggers, inadequate management, and sloppy status monitoring and reporting.

This is not an exact science. We are auditing a subset of the information and trying to extrapolate that to the overall portfolio, but if no problems are found in the sample that is selected for audit, then we can have a degree of confidence that there are no fundamental issues. Here again the organizational risk tolerance factors into the level of auditing that is performed. If there is a high tolerance for risk, then we may only audit 1% or 2% of the total risks. If there is a low tolerance, then we may be auditing 10% or more.

Process Improvement

As the owner of the organizational portfolio execution processes, the PMO is also responsible for delivering process improvement. This will occur in a number of ways:

- As a result of the process audit processes that we looked at above. This will result in a number of minor adjustments to the process that will solve specific weaknesses identified during the audits.
- From continuous improvement programs. All organizations should provide opportunities for process practitioners to make suggestions on how processes can be improved. They should also have processes in place that require process owners to review areas of the approach on a regular basis to identify adjustments that can improve effectiveness and efficiency.

- As part of a comprehensive process review. Processes will gradually become stale and outdated and will need to be overhauled in order to remain relevant and effective. This may be triggered by a major change within the organization or simply the growth and evolution that inevitably occurs over time.

Process improvement will need to address the changes to the processes themselves, as well as the tools and templates that are used by the process. However, it will also need to include updates to all of the support material—user guides and training courses, for example. Most importantly, the change needs to be implemented in such a way that the disruption to the portfolio is minimized.

This can be particularly difficult with organizational processes. If we wish to make changes to project risk management processes, then we can pilot those changes across one or two initiatives and monitor the success. We are minimizing the impact on the organization and are able to closely monitor the results of the changes, making adjustments on the fly if necessary.

At the portfolio level, this is much harder to achieve because the portfolio is both much larger in scale and is ongoing. For example, suppose that we wish to make a change to the risk analysis process. We can still pilot that change by asking just one or two of the people involved in risk analysis to use the new process while the others continue to use the existing approach. We can closely monitor those people to see how the revised process is working and what the individuals executing it think of the change compared to the previous approach. However, we are driving potential problems into the downstream organizational risk management processes.

Prioritization of risks, management decisions, contingency planning, monitoring, and reporting will now be based on a combination of different approaches. That not only makes it more difficult to compare risks with one another because of the two different approaches to analysis that have occurred, it starts to broaden the number of people impacted. No longer is it just one or two people conducting analysis, it is anyone who is making decisions based on the pool of risks. If the proposed improvement that is being piloted is ultimately not successful, then we could be driving problems throughout the organizational risk management environment.

That makes it vitally important that the PMO works with portfolio and program management in implementing any organizational process changes so that the impact can be understood and proactively managed, and this should be the case regardless of the type or scope of change.

The PMO remains the owner of the process; hence, the owner of process improvement, but change must be collaborative to be successful.

This partnership needs to start by aligning to the root cause of the problem—the real issue that is driving the symptoms that are being observed. The PMO will have conducted process audits and identified variances and trends, but the reasons behind those variances may be more complex. For example, if one particular project is experiencing problems with the effectiveness of risk identification, then there may be an issue of training or skills within the project team. There may also be an issue of a single project manager providing incorrect advice and guidance, or it may simply be that the project is complex and risk identification is not that easy. The full picture will only become clear when the PMO, the project manager, and potentially representation from the program and portfolio come together to analyze the situation and agree on the right resolution—training, supporting documentation enhancements, or maybe just some additional support in identifying the risks for a complicated project.

There also needs to be representation from all areas of the project execution process and the PMO in a continuous improvement program. This is particularly true where the continuous improvement approach involves elements of a staff suggestion process—where practitioners can submit suggestions on how the process can be improved.

Too often these initiatives lose credibility with staff because of a perception that ideas submitted to management levels are either ignored or rejected without explanation. By not only having broad representation from all sides, but also by having an open and transparent peer review process that involves the assessment of suggestions not just by management but by representatives of the people who actually execute the process, can some of this perception be removed.

There still needs to be a management level review and approval, and care needs to be taken to ensure that there is clear communication of decisions that do not align with the recommendations of the peer review process. The use of colleagues and practitioners not only gives these suggestion schemes more credibility, it can also lead to better solutions because they are being driven by people who are actually executing the processes, not the people who are monitoring them.

One of the most important functions that this PMO and project execution partnership has is in monitoring the results of process changes. Any change to processes needs to be monitored to ensure that the expected benefits are actually being achieved, but we need to consider those benefits from all sides. For example, if a change to a risk assessment process does not deliver the immediate gains that were expected, it could be because the changes did not improve the process. It may also be because it took time for the changes to become accepted, there was a ramp-up time for staff to learn how to make the most of the changes, the changes drove inefficiencies further downstream that now need to be addressed, or any number of other reasons. Only by looking at the change from all sides—PMO, portfolio, program, and project—will we truly be able to understand if further changes need to be made and what those changes are.

Independent Facilitator

In Chapter 8 on risk identification, we talked about the potential for a PMO resource to act as an independent facilitator. This is something that should be considered in a number of different situations and should not be restricted to specific roles or functions. There may be a number of occasions where an outside facilitator can be utilized, and the PMO should be the first area of the business that is considered to provide such a person.

While most meetings will benefit from some degree of facilitation, this is something that is generally carried out by the person chairing the meeting. An independent facilitator is usually only needed when the meeting chair needs to be deeply involved in the material of the meeting, such as may occur during a risk identification group review or similar meeting. This is an area where everyone involved in the portfolio or program needs to be able to contribute to the material being reviewed, and the discussions may become quite animated and detailed. By having independent facilitation, there is someone who is not a stakeholder in the specific outcomes of the meeting who can ensure that the process is being followed, that the ground rules for team and meeting behavior are being respected, and that the meeting remains on track and schedule.

There may be other situations where independent facilitation can add value, but these are more likely to be driven by a particular scenario—project teams that are experiencing frustrations and would benefit from an impartial person to help to defuse the tensions in meetings, for example. There are a number of advantages in having project execution experts, such as those located in the PMO, carrying out this facilitation role: they can relate to the team members, they understand the processes and challenges, and they will be respected by the people that they are facilitating. However, care needs to be taken to ensure that individuals performing a facilitation role do remain focused on the process and not get dragged into the material being discussed and debated. They may be in a position to add value to the material being discussed, but if that is to be their function, they should not also be the facilitator.

Expert Guide

A key PMO role that is frequently not leveraged as much as it should be is that of providing assistance to the project execution process. The PMO is frequently the place where the most project related knowledge is located, be that pure project management ability, organization specific project process experience, or just familiarity with the organization, its departments, and projects. This provides the PMO with a unique ability to provide impartial subject matter expertise and assist project teams with solving problems and overcoming challenges.

Often, however, that expertise is not leveraged. In many cases, that comes down to a fundamental lack of trust—a fear that asking for help is a sign of weakness, a suspicion that talking to the PMO about a specific problem will prompt an audit to be initiated, or a belief that the PMO will be unable to provide support. These all point to a problem with the organizational culture that we discussed earlier in this chapter and demonstrate that more work is necessary for that culture to become accepted and embraced.

When the PMO is able to act as an expert and provide guidance, care needs to be taken to ensure that the message is delivered and received in the right way. When a PMO expert is providing help and guidance to a project execution team member, the focus should be on helping that team member to solve the problem themselves, not on solving it for him or her.

For example, if a risk owner is unsure how to interpret a change that has occurred in the managed risk, then the expert from the PMO can help complete the analysis—identifying the questions that need to be asked (e.g., what has caused the change, is this likely to be a temporary blip or a permanent shift, is it isolated or could it be the start of a trend). They can also help the risk owner to identify other risk owners with similar risks who can provide some perspectives on the situation, and they can identify follow-up actions that can help to validate observations and decisions.

In doing this, the PMO is not only helping to solve the immediate problem but is also helping the project execution team member gain skills to solve problems in the future—a case of teaching someone to fish rather than giving the person a fish. This is the role all expert guides should take, whether within the PMO or not.

SECTION 3

17

Overview to Implementation

We started this book by looking at the concepts of organizational risk management, and then we looked in depth at the organizational risk management process and how it impacted different elements of the project execution approach within an organization. I've tried to provide some examples as we have gone along, but it's been based on a generic concept of an organization, and you aren't reading this book because you work for that generic organization. You work for a real company, authority, agency, or similar body. To benefit from this book you need to be able to apply these concepts in the real world.

That's what this final section of the book concerns—providing you with techniques to improve the quality of organizational risk management within your own work environment by converting the concepts discussed in Section 2 into a real process improvement initiative. This is a full-scale migration—a commitment from the top down in the organization to embrace an organizational risk management approach and implement all elements of it. However, not every organization can make that commitment, and it's not my intention to present this as an all-or-nothing solution. Adoption of any one of the organizational risk management elements introduced in this book can deliver benefits to the execution of projects in your portfolio. If you can demonstrate success with one process improvement on one element of one program, then you can start to leverage that across the rest of the organization. However, it should also be recognized that limiting the change to just (say) risk identification, without making changes to other risk related process elements, is not going to deliver the dramatic gains that can be achieved by a comprehensive strategic approach. Additionally, a partial solution will require additional integration steps with existing processes, which can create pockets of inefficiency as well as driving increased risk—rather ironic under the circumstances.

Do not think of these approaches to implementation in this section as an instruction manual, but consider them guidelines that can provide you with a planning template for your own implementation. You know your organization far better than I do, and you should trust your judgment on what will work, what needs more time and evidence of benefits to be widely accepted, and what just doesn't fit with your organization's project execution style. There are a few best practice sidebars at different points to try and reinforce some of the key messages, and they will help to focus you on some of the key success drivers.

Before we go any further, I also want to draw your attention to the Web Added Value™ Download Resource Center that is available at www.jrosspub.com. There you will find a number of downloadable templates and checklists to assist you in implementing organizational risk management. Again, view those elements as a framework and feel free to modify and adapt them to your specific situations.

Now, let's look at a few guidelines to bear in mind before we start looking at implementation in more detail.

It's a Project!

Have you ever noticed how bad project managers are at following formal project management processes for their own initiatives? It's as though the rules don't apply to them and that because they understand project management concepts they are somehow immune from the structure of a well-executed initiative. That's simply not true; a process transformation initiative, such as the rollout of organizational risk management, needs to be planned and managed just like any other project—a scope, a charter, key deliverables, and so on. If we are starting with a small-scale pilot of one or two processes, then we can forego some of the formality. However, if we are deploying a portfolio-wide organizational risk identification and assessment, then we need a manager, resources, milestones, and (most important of all) defined goals and objectives.

Throughout the book, I have been stressing the importance of achieving the goals and objectives of the portfolio, and a process improvement project is a great example of why that is the case. The physical deliverables of the project will be process flows, templates, tools, documentation, training programs, and sundry other documents, but none of those has any direct benefit for the organization whatsoever. The benefit only comes from the improvements that the application of those various deliverables drive into the way that the organization manages risks. In other words, the project (and this book) is not about the risk management process. It's about improving the quality of risk management within the organization through the development and application of effective processes.

The goals and objectives set for the project need to be realistic and highly visible. If the goals are completely unrealistic, then they will be rejected by the practitioners of the process, and the organization will believe that the processes don't work. For example, if we state that in the first year of organizational risk management, we will reduce the total number of risks that trigger by 75% and the number of unforeseen, management reserve impacting risks by 90%, we are setting ourselves up for failure. It's highly unlikely that we will achieve those targets, and that is going to rapidly become apparent to all involved. The result will be that the people executing the organizational risk management processes will feel they are wasting their time; they will either resent the new processes or consciously ignore them in favor of the previous way of doing things. This in turn will send a message to the executives that organizational risk management doesn't work, and any future expansion of the approach or extension to other organizational processes may well be cancelled. In reality, the only thing that was wrong was the unrealistic expectations.

Process improvement goals and objectives should also build over time. In the first iteration of a process, the people involved will be getting used to the new methods and templates. They will be learning what they have to do and how they need to interact with others; and they will be finding the insignificant problems that slow things down and need to be resolved. That needs to be reflected in the expected benefits for that iteratio, but by the second time the processes are executed there will be better familiarity with the tools and templates, better understanding as to how things work and what the process is designed to achieve, and the teething problems will have been ironed out. That requires the second iteration to have more aggressive goals—keeping the people who are executing the process focused on maximizing effectiveness and efficiency. By the third iteration things should be even better, and related process improvement projects that address other areas of organizational risk management should be delivering, which will allow for even greater benefits to be achieved.

This isn't an ever improving journey. There will come a point where the process is as effective and efficient as can reasonably be expected; any further improvements will be counterproductive—efficiency is lost in an effort to improve effectiveness beyond the point of diminishing return or effectiveness is lost by driving efficiency to the point where people are stretched too thin. While this will occur a period of months or years after the process improvement project is completed, this should still be considered at the time the project is executed because it is an important part of the goals and objectives.

The specific goals and objectives need to be set in conjunction with stakeholders, and they need to consider the needs and challenges of your organization. I would urge you to have a combination of organizational and practitioner goals. Examples of the organizational goals will be reductions in triggered risks, reduced risk exposure, and reduced utilization of management reserve (implying fewer unknown unknowns). Examples of practitioner goals will be increased process compliance and fewer process mistakes and omissions.

Implementing Risk Management Increases Risk

An ironic reality that we have to deal with is that the implementation of an organizational risk management approach will introduce additional risks to the rest of the portfolio—we'll actually be increasing the short-term risk exposure of the portfolio through our efforts to better control and manage risks. When we implement new processes, we are changing the way people work; that means moving them away from the well-established processes they are able to execute efficiently and effectively. Instead, we are presenting them with a brand new set of unfamiliar processes; this will result in them being slower to execute, more prone to making mistakes, and less comfortable with the work that they are doing, all of which will drive inefficiency and ineffectiveness. At the same time, the experience that people have with these new processes may potentially reduce their morale, further adversely affecting their productivity and level of engagement. This isn't a reflection on the processes themselves—rather, it is a natural human response when first faced with significant change.

When the change is the introduction of new processes in addition to the work that they are already doing—organizational risk management on top of project level risk management—things can be made worse. Workloads may increase; new employees may be added to the mix who not only don't know the processes, they also don't know the organization or the overall project execution framework; team members will be expected to take on additional responsibilities; and training on the processes will take longer than if it were simply a change to existing processes.

All of this has to happen against a portfolio of projects and programs that still need to be executed to deliver their own goals and objectives, and that's where the increased risk exposure really hits. New processes, new team members, increased workloads, and lower morale will all contribute to a reduction in the likelihood of success for the portfolio elements where the new processes are being implemented—the initiatives that have nothing to do with process improvement other than having been selected as pilots for the new projects. This should only be a temporary impact until the team becomes competent and comfortable with the new processes, but that doesn't make it any less real. Some of this can be addressed in the deployment of the processes, and we'll look at that later in this section. However, there is no getting away from the increased risk exposure, and by extension, the need for the organization to invest in additional reserves and/or additional risk management as secondary costs associated with the process development and deployment.

Commitment to the Work

As we discussed in Section 2, project related process improvement initiatives are generally considered the responsibility of the PMO. In many organizations, the PMO is not part of the project review and approval process dominated by more traditional business units—the various operational, development, and support organizations that represent the majority of project investments. As a result, PMO sponsored initiatives end up being *overlay* projects, initiatives that are conducted with any spare capacity resources and that are stopped and started based on the performance of other projects.

It should be obvious that this approach isn't going to work for an initiative and will fundamentally change the way that the organization executes on every initiative within the portfolio. In the next chapter, we will start looking at the concept of organizational readiness, but before we get to that stage, there needs to be a clear, conscious organizational commitment to invest a percentage of the organization's project budget to developing and implementing organizational risk management. We have said previously that the portfolio exists only to facilitate the achievement of the organization's goals and objectives. Therefore, the approved organizational risk management project should be tied to the achievement of those goals and objectives. That relationship should be fairly easy to demonstrate, regardless of what the specific goals and objectives are—an improved risk management process contributes to a greater likelihood of all other projects being successful, which in turn improves the likelihood of achieving organizational priorities.

An organizational risk management deployment project approved by the executive team as part of the portfolio of initiatives for the upcoming period will send a message to the organization as a whole that there is a serious commitment to driving improvements in project execution through better risk management. That commitment needs to be maintained throughout the execution of the work. Inevitably, there will be issues with some projects, and it's human nature that sponsors will try to protect their own initiatives at the expense of others. While you may not yet have a portfolio risk management process in place to help determine how to adjust the portfolio to deal with the problems and still retain the best possible chances of achieving the goals, the portfolio manager needs to ensure that the decision-making process is fair, neither protecting nor sacrificing the process improvement project unfairly.

Never Lose Sight of the Goals

Before we start looking at the various elements of developing and implementing organizational risk management within your organization, there is one final point to make in this introductory piece. When implementing a major process reengineering initiative, it's easy to become caught up in the details of the process—optimizing each step, addressing each detail of each template, and ensuring that all of the decision points are perfect. This is all important, but it can never be more important than the overall reason why the work is being done—in this case to reduce organizational risk exposure and improve organizational risk management.

We need to ensure that every decision we take in developing the processes and rolling them out is focused primarily on improving the effectiveness of risk management and secondarily on improving the efficiency of risk management. First, we make it as good as it can be, and then we minimize the cost without compromising how well it works. That's never an absolute measure. As we talked about in Section 1, the organization is constantly shifting and evolving; that shifting will change the effectiveness and efficiencies of the process on an ongoing basis, requiring adjustments to be made.

At the same time, the project execution maturity of the organization will be growing, and there will be an ability to implement more complex processes that would previously have been overwhelming. This will lead to a need to review and revise processes that have already been deployed as we implement new elements that add to the overall methodology. For example, organizational risk assessment may drive changes into a previously implemented organizational risk identification process in order to ensure that the handoff between the two elements is aligned and consistent.

This should be expected and is a sign of a maturing and evolving process, but every time that a change is made, there needs to be a conscious check that the overall effectiveness of the process is being enhanced or maintained. We should also be looking to improve or maintain efficiency; it is acceptable, however, to compromise efficiency for the sake of effectiveness, if only until the efficiency can be restored—there may be a cost associated with reducing the organizational risk exposure.

Go into this work with eyes wide open. If you and your organization start down the path of an organizational risk management implementation, you are making a commitment that is likely to gain tremendous momentum. The increasing maturity will not only drive growing complexity in risk management, it will also drive a desire for other portfolio level processes. While you may have some degree of portfolio level cost and resource management, you likely don't have anything for quality and may not have even considered things like communication, change, or issue management at the portfolio level. This is all incredibly positive, but it will require a commitment to an ongoing evolution of the organizational project execution methodology that will expand beyond process and into enterprise software tools, organizational structure, hiring and development, and virtually any other element of the portfolio.

If that sounds like an exciting prospect, then let's get started!

Web
Added
Value™

18

Organizational Analysis

Implementing a major process doesn't start with the drawing of process flows, or even with the creation of the project plan. Instead the work has to start with understanding the current state and the amount of change that the organization can withstand without impacting the ability to execute the portfolio. This work is part of business case development; however, with the need to investigate a potentially broad spread of departments within the organization and to look at areas that departments may consider sensitive, there may need to be a preliminary research project approved to provide the authority that is required for the analysis to be successfully completed.

In this chapter, we look at some of the different variables that are in play and consider how they might impact implementation speed, approach, and even the ultimate solution. Only you know how your organization rates against these different criteria, and you will likely have additional measures to consider. That's okay. Success can only come if you have a customized solution for your own unique environment.

Once this organizational analysis is complete, you should have a solid proposal/business case that can be submitted as a project candidate for approval and inclusion within the portfolio of initiatives the organization is going to undertake. The groundwork completed in this section will help to ensure not only that you have a proposal that is approved, but that the scope of the initiative is the best fit for the organization's needs.

Portfolio Management Maturity

There is a fundamental question that needs to be answered before any kind of organizational risk management approach is implemented: *is my organization ready for this process*? Most organizations are now embracing project management to a greater extent, and for many there is also recognition of the benefits that program management can deliver. However, when it comes to portfolio management, there is still a lot of disparity between organizations in terms of how it is implemented, or even whether it is implemented at all.

I still see a lot of organizations who consider the portfolio just once a year—when the annual planning and project approval cycle occurs. After that, the work is distributed out to different departments and managed in accordance with the processes existing within those departments and the group specific PMO (or less formally the organizational group of project managers). This may sometimes represent a program driven culture, but clearly, this is not an organizational model that is yet ready for a single, organization-wide, risk management approach based on the fundamental concept of a portfolio manager who is responsible for delivering the organizational goals and objectives represented by the portfolio and its constituents.

That doesn't mean that some degree of organizational risk management cannot be successful. Rather, it means that the approach has to be tailored to the level of maturity that the organization has. In this particular scenario, I would recommend finding a more progressive, forward thinking PMO or department within the organization and implementing a program level risk management approach for one of that group's programs. It can serve as a pilot initiative for other areas of the organization. This will help to contain the impact and will also minimize the potential for resistance to change. If the organization hasn't yet embraced portfolio management as a concept, then there are likely a number of people who are resistant to changing existing processes that assume a portfolio management structure.

At the same time, by implementing an organizational risk management approach within a program, we are creating an opportunity for comparison between that program and others. If other departments, program, or project managers see opportunities to improve their own initiatives based on the results from the pilot, then we start to create a foundation for future expansion. We begin to create interest and enthusiasm in the concept, and, more importantly, we start to enhance the organizational portfolio execution maturity.

If the organization sees the benefits of an organizational approach to risk management, either based on the results of a limited pilot or from a recognition that the concepts will drive value, this should be encouraged, even if the organization is not yet mature enough for a portfolio-wide risk management process. In this situation, there needs to be a larger scale program that will address the formalization of a portfolio based structure and execution approach. Risk management will be a key part of that work, but it cannot be the only piece.

Process Environment and Culture

When we have considered portfolio management maturity, we have to look at the processes that exist within the organization and the culture around their use. This isn't just risk management processes; here we will consider all of the project execution related processes, as well as going beyond the processes themselves to consider:

- Attitude toward process change—are processes updated on a regular basis, or are they hardly ever changed? If the organization is used to changes being made to project processes on a regular basis, they are more likely to be receptive to change than if the project execution processes have remained unchanged for several years.
- Success of process enhancements—is there a track record of implementing processes that have been well thought out and tested, or is it more common for a process to go through two or three iterations before they are effective? If there is frequently a need to go through multiple versions before the process is successful, there is likely to be some initial skepticism about a brand new process.
- Approach to process change—are changes generally implemented as a slight adjustment to a group of processes or a comprehensive overhaul of a single process area? If the organization is used to only slight *tweaks*, it will be harder to implement a major new process then if there is familiarity with major enhancements.
- Attitude toward existing processes—are they embraced and accepted, grudgingly followed, or openly ignored? If project compliance is a problem within the organization, there may be a need to address the underlying causes of that before implementing new processes. On the other hand, if the practitioners have been proclaiming the need for process upgrades, change may come more easily.
- Relationship between the PMO that owns the process and the practitioners that apply the process—is there a constructive partnership or an adversarial relationship with mutual distrust? Unless the two sides are working together, there will be additional barriers to implementing process changes.

While none of these aspects have any direct connection to the organizational risk management process itself, or on any other individual process or consolidated methodology, they combine to create the environment in which processes have to succeed. Any one of them can make or break the success of the implementation, and there needs to be recognition that each of these elements must be considered in developing an implementation plan for the new processes.

Of course, we also have to consider any processes already in place as part of portfolio and program execution. Even if the organization is mature enough from a portfolio level to have portfolio execution processes in place, we need to consider their appropriateness for integration with organizational risk management. If there are only rudimentary processes with no detailed documentation, templates, or commitment to compliance, then we probably don't want to spend much time determining how to integrate organizational risk management with them. At the same time, we can't simply leave things as they are and have organizational risk management as a separate stand-alone process that is completely independent from other portfolio or program processes.

Even if the current commitment is only to implement a pilot organizational risk management approach, we need to understand what the potential *end game* is. If success can lead to the development and implementation of an enterprise-wide portfolio management methodology, then let's architect the risk management process for one department with that framework in mind:

- Ability to scale—ensuring that the process approach we propose will be capable of expansion into different areas of the business and that it can support significantly higher volumes of risks than are anticipated in the first iteration. This may mean avoiding using technology tools only available to part of the organization, avoiding a reliance on too many manually driven processes, and even something as simple as avoiding the use of department specific terms and acronyms.
- Ability to integrate—building the risk management approach with the understanding that it may need to integrate into a group of additional processes to form part of a larger portfolio execution methodology. This will mean considering hand-off points, interaction with other processes, and inputs or outputs from and to other processes. It may even consider things like making sure that standard corporate template structures are used to avoid having to recreate the work.
- Ability to leverage—undertaking the work and documenting it in such a way that the same approach can be used to develop additional processes through later projects or phases. The most

obvious example will be to ensure that the way that the project proposal is developed is well defined and documented so that it can be used to assist in the development of future proposals.

These items assume that organizational risk management is the first such process to be implemented. However, there is also the possibility that work has already been done on other organizational processes. If that is the case, the proposal and framework for the risk process should seek to leverage that work wherever possible—consistent terminology, template styles, etc.

Risk Management Success

In recent years, there has been a tremendous increase in the utilization of Agile techniques for project execution. While there have been many reasons for that growth, a lot of the success has been nothing more complicated than the fact that organizations were experiencing high failure rates with traditional waterfall based project methodologies and were ready to embrace something different. Agile came along at the right time and proved to be tremendously effective at delivering the improvements organizations desired. If organizations had been experiencing high rates of success with waterfall based approaches, then there would have been a slower uptake of Agile—it would have been just as effective but wouldn't have been needed as badly.

We can apply that analogy to organizational risk management. If organizations are experiencing high levels of project success by doing things the way they currently do, then there will be less incentive for them to change and look for new ways of doing things—the belief will be that if it's not broken, it doesn't need fixing. This may in fact be one of the reasons why an organization has been slow to adopt more formal portfolio management techniques—there isn't a perceived need for it.

This may truly be the case. An organization's risk management approach may have been implemented in such a way that the program and portfolio level risks are identified, analyzed, and managed without any formal organizational level approach. However, I think that we can all recognize that would be a rare scenario. It's more likely that the organization has no appreciation for how much better things could be with an improved approach.

Let's look back at the comparison with Agile. When organizations moved some of their software development projects to Agile, they saw immediate and dramatic improvements—improvements that had a material impact on the organization's overall performance—fewer defects, better customer engagement, solutions that better met customer needs, and of course faster time to market. This made for good reading, and the early successes with Agile became well publicized, leaving CIOs with minimal doubt that they needed to explore Agile concepts or be left behind. At the same time, product owners were complaining about losing market share to competitors who were suddenly delivering solutions that better met the market's needs with fewer defects and shorter timelines.

With portfolio management in general and risk management in particular, there isn't such obvious visibility. There is no easy comparison of the quality of risk management in your organization with industry norms, no ability to tie a competitor's improved performance back to changes that they have made in risk management, and no normal performance to compare our own risk management to in order to determine whether it truly is world class. Instead, we are left with our organization's own perception of whether the current risk management approach is successful. In many cases, what is perceived as *good enough* may simply be a situation where there is no understanding that things could be significantly better with an organizational approach.

Risk Awareness

The old cliché is that you can't manage what you can't measure, and that's true, but in the case of risks, it doesn't seem to make any difference. Organizations frequently fail to identify all potential risks and fail to properly plan for the risks they do identify. If your organization is still consistently failing to identify all of its risks, then there is going to be minimal appetite for implementing any form of organizational risk management. That will force the organization to acknowledge risks it has previously decided to ignore (consciously or otherwise), which in turn will drive a lot of changes to the project execution approach, beyond just risk management.

A large part of the problem is that many organizations still think of projects in terms of the traditional triple constraints model—cost, schedule, and scope. Risk is immediately relegated to a secondary consideration at best and risk management activities are viewed through the lens of the triple constraint—effort and money spent managing risks, increased contingency and management reserves, and few or no project deliverables. Frequently, risk management continues to be an exercise undertaken during project planning—identification and analysis combined with nominal ownership assignments. The number of organizations that still manage risks using the blind optimism approach is shocking.

Even when risks trigger, they aren't recognized as such; instead, they are new issues that have suddenly cropped up and need to be dealt with— a team member resigns, a vendor delivers late, or a deliverable fails quality control checks. If the organization doesn't recognize that the effort that could have been expended to reduce the likelihood of any of these events from occurring, or preparing for their impact, would have been less than the impact that the events have caused, there is unlikely to be any recognition that there is a need for improved risk management.

It is not uncommon for project sponsors to refuse to acknowledge the need for contingency reserves, let alone the less well-defined management reserves, generally because it means that less project scope can be delivered for the same investment, or that more investment is needed for the required deliverables. Of course, refusing to accept the need for the reserves doesn't remove the need for them, and the impact is simply that projects have to try and absorb the variances that triggered risks cause. Unfortunately for many sponsors, this is still considered to be a better solution—generally an indication that the organization is still conducting projects on a distributed model with control in the hands of department heads rather than a central portfolio based approach.

Organizational Constraints Hierarchy

We have discussed the concept of the constraints hierarchy earlier in the book; clearly, it is a major influencer on the readiness for an organizational risk management process. The organizational constraints hierarchy is somewhat fluid, but there are some frameworks that we can establish for our organizations. For example, it's fair to say that a company involved in the development of new drugs is always going to have risk at or close to the top of the constraints hierarchy, especially when it comes to risks that directly impact the quality of the drugs under development—the impact of a triggered risk may literally be life and death.

On the other hand, a company involved with the development of web or mobile applications may not have risk as high in the constraints hierarchy. For them, time to market (schedule) and cost may be more important, and they might be more willing to accept risks because they can recover with relatively low impact through patch releases or new versions. That's not always going to be the case, especially where data privacy and security are concerned, but generally speaking, the impact of risk is less significant for these companies.

You may think I am going to suggest that an organizational risk management process is more important for the drug developer than for the online software company, but that's not the case. I consider both situations as important for the implementation of a process, but the focus of that process will be different. In the case of the drug company, they need an organizational risk management approach concentrated on identifying and managing the key product related risks as effectively as possible. Their entire focus is on minimizing the number and impact of triggered risks with the impact on cost and schedule being secondary concerns.

Best Practice—Making Sure Your Organization is Ready

A successful project requires realistic goals. Organizational risk management can only be successful if the solution is designed to meet your organization's needs today; if you try and implement a Ferrari solution in a Model T organization, then bad things will happen!

To maximize the chances of success, make sure that in preparing your proposal you consider the following questions:

- Am I proposing something that portfolio, program, and project level resources will be able to execute with their current skill and experience levels?
- Will organizational leadership recognize the need for organizational risk management and commit to it?
- Are existing processes robust enough to work with this kind of change?
- Do we understand enough about risk management as an organization to make the proposed approach work?
- Do we care enough about risk management as an organization to make the proposed approach work?
- Is my proposal the right one for this organization?
- Do I have the right supporters?
- Am I proposing this at the right time given everything else that the organization is trying to achieve?
- Is this going to make the organization more effective at executing the portfolio?
- Is this proposal as good as I can make it?

If you can't answer yes to every one of the above questions, then you may need to do a little more work on the proposal before it is ready to be submitted for review.

In the case of the software company, they need a process that will allow them to quickly determine what is likely to happen if the amount of effort and/or money invested in risk management is changed. It also needs to identify the risks that can have reductions to risk management with minimal impact on the overall initiative so that they can divert resources to schedule driving (critical path) tasks if necessary. That requires a process that has accurate and detailed risk analysis and prioritization elements so that there is confidence that when compromises have to be made, the right risks are being accepted in order to preserve the schedule and budget.

Selecting Champions

Successfully implementing a major process such as organizational risk management requires the alignment of a number of different roles. First, someone has to come up with the idea for the process and be prepared to work to develop and implement the concept. If you are reading this book, then that may well be you, regardless of the formal role that you have in the organization. Second, there needs to be alignment with the PMO(s) or EPMO that will ultimately own the process when it becomes part of the project execution methodology. Clearly, the PMO also needs to be in agreement with the idea that the process is needed; if the PMO doesn't support the concept of the new process, there is little chance for success.

There is also a third role that needs to be considered—that of the sponsor. In many projects, the sponsor is obvious. It's the head of the department executing the project, but in the case of a process improvement initiative, there is more flexibility. The equivalent to the department head will be the head of the PMO—the PMO will ultimately own the process and so is a natural selection to champion and promote the initiative during the project selection process, as well as being the driving force behind the initiative during execution. However, there may be other candidates for the role of sponsor who have more potential impact within the organization.

In the portfolio management maturity section above, we used the example of prototyping a process within a single program. The program manager acting as an initiative sponsor sends a different message than if it were the PMO—the person responsible for the program is saying that they think that this work can help them. A similar, and potentially more powerful, message is sent if the program sponsor or customer supports the process improvement initiative.

During the organizational analysis, you need to establish who the most appropriate champions will be—people who are going to see the benefits of the changes and will be prepared to drive them forward within the organization, and people who are senior or important enough within the organization to deliver a message that this work is important. You need to recognize that there may be a need to compromise some elements of the process in order to win that support—politics is a reality of organizational life, and unless your vision of the ideal organizational risk management approach perfectly aligns with that of your preferred sponsor, you may need to adjust the approach to achieve overall success.

That compromise may simply be a slightly heavier emphasis on an aspect of the process that is more important to the sponsor, using one of the sponsor's risk management approaches as a base, or using some of their resources in your team. If you find yourself having to compromise a fundamental element of your idea, then you may not have the right champion for your initiative. On the other hand, if you find a sponsor that accepts your proposal without modification, then you may find there is not enough engagement with the initiative—it's unlikely that you and your sponsor are perfectly aligned. This can become a complex balancing act, which is why it is important to conduct it during this initial analysis phase to avoid having to make sudden shifts partway through the process.

Organizational Priorities

As we discussed earlier in the book, the organization will establish its strategic priorities and then approve projects as the vehicles to try and achieve those priorities. When project candidates are being considered for inclusion in the current portfolio, the match between a proposed project and the corporate priorities is going to have a large say in determining whether the project is approved.

Therefore, it follows that if the implementation of organizational risk management helps an organization to achieve those priorities, it is more likely to be approved than if it is incidental to those corporate priorities. We looked at this to some extent when we talked about the alignment with the organizational constraints hierarchy, but this is more fundamental. If organizational risk management is implemented well, then it should always improve the chances of the organization achieving its goals (after all, that's risk management), but we have to consider positioning of the proposed process, as well as the process itself.

For example, suppose that an organization's top priority is to reduce costs by $x\%$. An effective organizational risk management process should contribute to cost reduction in a number of different ways:

- Better risk identification will lead to fewer *missed* risks, which will mean there are fewer issues for which we are unprepared; hence, less time and money is wasted identifying the extent of the problem and preparing to respond.
- Better risk analysis and prioritization will lead to more appropriate allocation of risk management resources where they can have the best return on investment.
- Better risk management will lead to reduced risk exposure, better contingency, and reduced needs for reserves.
- Better impact assessments will lead to improved ability to understand the scope of impact, faster recovery, and more appropriate lessons learned that can be converted into reduced future risk exposure.

These are the elements that a proposed organizational risk management initiative should be emphasizing in this scenario—the elements of risk management that specifically speak to how organizational risk management will support the desire to reduce costs. If the business case focuses instead on how the organization can increase revenue through quicker delivery, improved quality, and a better fit with requirements, then stakeholders are less likely to approve the initiative because there is less obvious alignment with the corporate goals.

This may seem shortsighted. After all, an organizational approach to risk management will deliver benefits for many years to come; in that period, the organizational priorities will evolve and shift. However, the decision on whether to invest in developing and implementing the approach is going to require funds from the current portfolio budget. Unless there is some immediate contribution to the current goals and objectives, it will be difficult for an initiative to gain approval. The benefits may be clear, but few executives will be prepared to sacrifice the ability to achieve their own targets for the greater good of the organization. For the same reason, the business case should seek to deliver some in-year benefits—it's a lot easier to commit to a project knowing that there is going to be some kind of return within the reporting period.

Organizational Needs

When you are conducting the organizational analysis and developing the business case for an organizational risk management initiative, there should be one overriding question that you ask yourself as you consider all of the elements outlined in this chapter: how will the proposed solution improve the organization?

We can never lose sight of the reasons for implementing this or any other type of process—it has to be making things better for the organization. Sometimes *better* is an easy concept to grasp—doing the same thing for a lower cost, generating more revenue without incurring additional costs, etc., and this aligns with the concept of the portfolio goals and objectives. Sometimes the benefit is less obvious, and people struggle to connect the work to the benefit. If a project to implement organizational risk management is approved, the costs are easy to see—we spend effort, time, and money on developing the processes; we divert resources from other initiatives that would otherwise have been approved; and we disrupt the way that projects are executed.

We have to make sure that the benefits are just as obvious—a 5% reduction in risk exposure may be a realistic claim, and we may be able to demonstrate that kind of improvement after the process is improved, but that's not a concept that is easy for executives to grasp. Instead, if we position the benefit as freeing up $1 million dollars for investment in the portfolio (the equivalent of a 5% reduction in risk exposure on a portfolio where the combined contingency and management reserves were $20 million), then that's suddenly a different conversation, and a much easier sale to make. That's a slight oversimplification because there will be some costs associated with the operation of the process, but the concept is clear—tie the benefits to the organizational needs.

This also takes the benefit beyond a simple alignment with the goals and objectives. As noted above, those will change and evolve over time (sometimes even within the year that is being planned for), and the claimed alignment may not actually occur in reality. However, the benefits from a reduction in risk exposure will always deliver tangible and obvious benefits to the organization and are an ongoing benefit year after year, not just a one-time lift.

That theme has to run through the entire proposal or business case that you present. The process is not being proposed simply because it's a good idea. It is being done because it will result in improved financial performance, regulatory compliance, customer satisfaction, and/or organizational capability through the better management and control of risks.

Leveraging the Analysis

All of the elements considered in this chapter will have a direct impact on the nature of the project that is proposed, the way that it is positioned within the organization, the scope of the work, the likelihood of the initiative being approved, and the reaction of stakeholders to the planned work. If undertaken properly, the analysis should result in a business case or project proposal that has a strong chance of being approved. More importantly, it should define a project that delivers significant improvements to the way that organizational risk management is conducted within the organization.

Reading the sections in this chapter, it is easy to see that there are a number of connections between areas—priorities and the constraints hierarchy, needs and the right champion, awareness and success. The analysis needs to treat all of these sections as part of a larger project execution ecosystem with each of the elements impacting and being impacted by the others. If you focus on only one or two areas, then you are likely to find that your proposal is incomplete. Even if it is approved, you will run into problems during the process design and deployment work due to incomplete or inaccurate scope, untested assumptions, or unidentified constraints.

It should be obvious by this point in the book, but an incomplete analysis will also result in increased risk exposure in the event that the initiative is approved.

19

Project Initiation

Congratulations, your business case was approved! Now you actually have to develop and implement the process. I'm going to assume that if you have made it this far, you either have enough project management skills and experience to not need me to tell you how to plan a project, or you have access to people who know how to do that for you. Therefore, I'm going to ignore things like developing work breakdown structures, estimating work, building critical paths, and developing resource assignments. I still think that work is vitally important if you are going to use a waterfall based approach to project execution, but I don't think that it's necessary to review those items here.

If your organization has adopted Agile approaches and is comfortable with them, then you may consider that for this process development work. Just because the deliverable of the project is a process rather than a piece of software doesn't immediately exclude Agile from consideration. In many ways the structure of short sprints and frequent customer feedback loops works well with the development of a process, as long as the output of each sprint is self-contained enough to be meaningful to the group reviewing the work in the role of the customer.

Where the work outlined in this and following chapters does wander toward project structure (and inevitably it will), I am going to assume a traditional or waterfall based project model. That is simply because I anticipate that is how most people will go about such a project; it is not intended to imply that other project structures won't work.

The Right Start

As I said at the start of this section of the book, I am going to assume your organization has approved a comprehensive top-down initiative to implement all elements of organizational risk management. If your organization approves a different scope, then you can adapt this approach to fit that scope; the concepts will be the same. Before you do anything, you may need to determine whether you have a project or a program. This might not have been determined yet, even though the initiative has been approved. If there is going to be a phased approach to the rollout, then a program may be appropriate, even if there are conditions that have to be met before the later elements are approved. If the organization is large and/or has distributed PMO structures, then a program approach will almost certainly be beneficial. There is also the possibility that the organizational risk management initiative is part of a larger portfolio methodology deployment, in which case organizational risk management may be a project (or series of projects) within a portfolio execution methodology program. On the other hand, if this is a single piece of process development work with no immediate plans to expand the work into other organizational areas or other processes, then a simple project structure will be perfectly adequate.

Regardless of whether this is a project or a program, before the work starts we have to determine how we are going to structure the initiative and how we are going to go about managing the work within the group. Some of this may be defined for us in the business case. The rest will be determined by the team and captured in the project charter, project approach document, and various project plan elements that should rapidly follow the approval of the project. A few fundamental items need to be considered:

- *Identification of stakeholders.* Your major stakeholders will be fairly easy to identify, but success depends on effective management of all stakeholders; you can't do that unless you have identified all of them.
- *Sourcing of resources.* Anyone who has managed a project knows that you don't always get the resources that you want, but with process improvement in particular, it's important to get the right mix of people.
- *Communication strategy.* A lot of people will be interested and concerned with the work that you are planning to undertake, and you need to ensure that you are giving all of them the right information, at the right time, in the right way.

- *Organizational integration.* You need to consider how organizational risk management will align with the other processes that are in use within the organization and how handoffs will occur.

Let's look at each of these individually.

Identification of Stakeholders

The definition of a stakeholder generally identifies three distinct functions or roles that a stakeholder may perform:

1. Anyone who is actively involved in the project will be a stakeholder—team members, the customer and sponsor, and everyone in between.
2. Anyone who may be impacted by the work of the project is a stakeholder.
3. Anyone with the potential ability to impact the project (work or people) is a stakeholder.

Based on that, you are going to be busy—everyone involved in the execution of the portfolio or an element of the portfolio is going to be a stakeholder, so too are all of the sponsors of those portfolio elements, along with resource owners and PMOs. If we take a fairly broad interpretation to point number 2, then regulators, shareholders, and customers can also be considered stakeholders.

That may sound a little ridiculous, but it's actually a good perspective to have. We should be developing our processes with the needs of all stakeholders in mind; that means that we need to fulfill the demands of not just the people directly associated with the project, but also the demands of the organization as a whole. We do need to ensure we are contributing to the achievement of the goals and objectives that our shareholders will measure us against, the quality of the products that define how customers view us, and the compliance with regulations that will be the concern of regulators. If we aren't delivering solutions that improve each of those areas, then we aren't doing our job correctly.

Of course there is a difference between these stakeholders and the team members, resource owners, sponsor and governance committee that we think of as more traditional stakeholders; and this requires us to categorize our stakeholders rather than simply form a long list. Broadly speaking, I tend to think of stakeholders as belonging to one of three categories:

- *Active stakeholders.* These are the people who are actively involved in the execution of the project—they are our team members, the owners of resources, the sponsor, etc. You may hear these referred to as core stakeholders in some organizations, and they also represent what many people are referring to when they think of stakeholders in a narrower context.

- *Passive stakeholders.* These are the people that we still need to interact with, but who are not actively engaged in the work of the project. These are usually the people we keep informed of progress; in the case of our risk management initiative, the project managers who will ultimately use the process are a good example.

- *Detached stakeholders.* This is the group of people that fall into that broader definition and whom we will likely not be directly connected with during the execution of the project, or where we will only communicate with them occasionally. However, we should always be asking ourselves whether the needs of these stakeholders are being met—this group can provide a valuable role on the project by helping us to remain focused on the ultimate goals of the initiative, if only as an abstract concept (what would the stakeholders want us to do?). Remember, we aren't trying to implement a better process; we are trying to deliver improved results through the execution of a better process.

Sourcing of Resources

Portfolio level resource management is an extremely valuable approach to an organization. It looks at the portfolio as a whole and determines how best to distribute the skills and expertise in order to maximize the return on the investment. If your organization has a well-established, well-executed portfolio level resource management approach, then count yourself lucky and make the most of it. You will likely end up with the best mix of resources to meet your goals given the multiple demands on the limited resources available.

Unfortunately for many organizations, resource assignment is still a competition between line managers and project managers and between one project manager and the next. You can't allow yourself to become embroiled in that kind of discussion; it will only ever lead to distrust and resentment with some of your stakeholders. Instead, you need to recognize that you won't always get your first choice resources—there are many other projects competing for resources as well as operational needs, and the organization as a whole needs to ensure that each person is assigned to a role that will maximize the return on the investment in that resource.

Instead you need to focus on your project's specific *must have* needs—the elements of your initiative that just can't be compromised.

This shouldn't be focused on individuals, but rather on skills—it's not that you have to have a certain person; it's that only that person has the skills you need in a specific area. The details will be unique to your own project, but these areas need to be considered:

- *Single points of expertise.* Is there a role on your project that requires skills that only one person has? That's not necessarily going to be a common situation on a process development initiative, but there may be someone who was responsible for writing the current risk management process, and they would be a significant addition to the project.
- *Resource mix.* Process development is not an academic exercise—in this case, it's going to result in an approach to risk management that will need to be executed in the real world of projects within the organization. We need to ensure that the team comprises a mix of process and project experts, people who can help ensure the process is effective and efficient, as well as people who can help ensure the process can be applied in real projects, programs, and portfolios without falling apart or acting as a drag on the other work elements.
- *Department mix.* One of the biggest problems that process improvement initiatives face is rejection of the process—practitioners refuse to accept the new processes and continue to work in the ways they always have. To try and counter this we need to consider having representation from as many of the impacted teams as possible—it's a lot harder to reject something you had a hand in developing. Even if the process is initially going to be deployed in just one of the organization's PMOs, consider having representation from all PMOs, so it doesn't appear to be a solution tailored for just one part of the business if and when it is expanded throughout the organization.
- *Personality mix.* It's always important to have a group of people that can come together and work as a cohesive unit, but in process development, a lot of the work will involve working in groups—discussing different proposed process approaches, capturing needs and presenting solutions to different parts of the organization, and working through the process details of each element. This requires a much higher level of group engagement than many projects, and you need to ensure you have people who are comfortable working in group environments.

Of course there is also the need for all of the skills and attitude that any project should have—people who are committed to the initiative, believe in its goals, understand the importance of deadlines, can work on multiple items at the same time, and are used to the project tracking and management infrastructure. Just like any other project, and indeed the entire organization, the team should have a mix of youth and experience, senior and junior staff, high performers, and those who need to develop.

One common experience with process development projects is that many resources are only assigned part time. Creating or enhancing process is frequently not seen by resource owners (or even the organization) to be a full-time role for individuals, and this may result in the project manager finding that they are only given resources for a percentage of their time—even if the project budget allows for a higher percentage allocation. This isn't necessarily a problem. There may even be advantages to having people working on other initiatives at the same time because it keeps their risk management awareness and project experience fresh.

However, there needs to be recognition among resource owners (and project managers when they are procuring people) that this is not a part-time project. The work is not in defining the process and documenting it. It is in the process of determining what the right process is, how it needs to integrate with other processes, how it can be maximally effective, how it should be introduced and rolled out, how people should be trained, how its success should be measured, and how it should grow and evolve. People are not being assigned to the project to produce specific deliverables, as they might be in a construction or software development project. Instead, they are being assigned because they have the right skills and experience, and most importantly, the ability to apply those skills and that experience into determining the best possible organizational risk management process.

Communication Strategy

Every project manager knows that communication is 90% of the job, with everything else having to be completed in the other 90%! I have no intention of telling you how to communicate with your teams and stakeholders on the issues surrounding the project, but there are some communication challenges unique to process development initiatives that need to be considered.

The problem with developing a new or changed process is that it involves changes to the way people perform their work, and human nature is such that many people will assume the change will have a negative impact on them. When the process being changed is at the organizational level, there will be an impact on a greater number of people; hence, more people will fear that their work will become more difficult, that they will be under closer scrutiny, and that their decisions will be second-guessed. When we go further and address risk management processes, there is an even greater concern among employees that things are going to get worse. Anyone who has ever worked on a project will be able to provide examples of situations where they were blamed when a risk triggered or was missed entirely. The fear with a new risk management process is that it will just result in more of the same.

This means that there has to be specific communications to stakeholders on the contents of the new process, especially to the group that we defined as passive stakeholders above. This is more than simply status reports that provide brief summaries of what's going on, and it's more than involving those stakeholders in reviews and testing (although that's important). Rather, it's about proactively creating the right environment for the process to be deployed—preparing the people who will be impacted for the arrival of the process.

Like any communication, the key is to provide clarity and consistency in the message. That means that you need to consider what you are going to say, when you are going to say it, and how you are going to deliver it. You want to avoid a perception of keeping people in the dark, but you also want to avoid over communicating, which will just cause people to stop listening. Similarly, you want to make sure you keep people informed of what is happening while avoiding meaningless updates.

This means making sure that the passive stakeholders fully understand why the process is being developed, the challenges that we are seeking to overcome, the opportunities that we want to exploit, and the enhancements that we need to make. It also requires us to be clear on the scope of the process. If we are implementing it initially in just a single part of the organization, then make that clear. However, if the intention is to take the approach organization wide if it is successful, then make that clear too. Transparency helps to develop trust that people are aware of what is really happening.

We should expect and encourage questions and suggestions. Not only will this help us to understand the areas of concern, and by extension the areas that we should focus on in our communications, it provides us with input from other stakeholders on how to improve the process that is being developed.

Organizational Integration

This concept is closely associated with the communication section above but is specifically focused on how the process will change other elements of organizational process. The organization will have established ways of doing things, whether they are formal processes, informal guidelines, or simply undocumented habits that have become accepted over an extended period. Even if none of these has any direct overlap with risk management, the organizational risk management process we implement will impact some of them:

- Documents that become inputs to the risk management process may change with the addition of new data elements that the risk process needs.
- Outputs of the risk management process may form new or modified inputs to existing organizational processes, which in turn may also drive changes to the process elements in order to incorporate the changes.
- New organizational tools may be required—the organizational risk profile and organizational constraints hierarchy are perfect examples. These aren't directly part of the risk management process, but they need to be either included in the project, or in a dependent project.

This last one in particular goes quite a long way beyond the scope of risk management, but it is vitally important work, and we need to determine how it is going to be addressed. The simplest solution is to include this work within the scope of the project, or as an element of a larger program if there is an initiative underway to implement an organization-wide portfolio execution methodology. However, if this is not yet an organization-wide initiative, things may be more complex.

Tools such as the organizational risk profile are likely going to be developed initially because of this initiative, so there is a natural fit with the project. However, other processes, inputs, and outputs are likely to be well established already and working smoothly within the rest of the organization's process infrastructure. If we need to make modifications for the sake of organizational risk management that are not required for the other areas where these elements are used, we need to find a way to minimize the impact while maximizing the chances of success for our project.

Many organizations will decide not to make changes to existing processes or artifacts if the initial deployment is simply a pilot or is limited because if portfolio level risk management is not ultimately deployed throughout all areas of project execution, then the changes will be unnecessary. The problem with this is that it can be a self-fulfilling prophecy—the reluctance to optimize processes and/or documentation to align with the needs of organizational risk management is likely to reduce the effectiveness of risk management, which in turn will reduce the likelihood of the project being approved on an organization-wide basis.

My preference in situations of potential conflict between new and existing processes is to build an additional integration point into the process, similar to that shown in Figure 19.1. Here you can see that we leave the existing processes, inputs, and outputs untouched and add a modifying process within organizational risk management. This is simply a review and validation step where any additional data is added to templates and any further interpretation of data or similar activities can occur. This isn't a perfect solution; it will likely be less efficient because it requires an additional step to be performed. It may also be less effective because it is one step further removed from where the work element is created or modified. However, it represents a compromise that allows for existing processes to continue unaffected while still allowing the organizational risk management process to be more effective than if there was no integration.

Note that some of the arrows flow in two directions because it is possible that this process will be two way—the risk management process

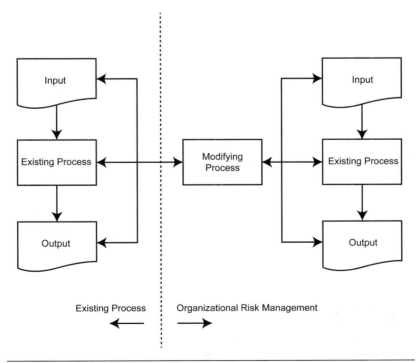

Figure 19.1 Process integration approach

may modify items that are needed within existing processes, as well as the other way around. Not every arrow is bidirectional because it doesn't logically make sense—items won't flow directly into the outputs of a process or out of the inputs to a process.

This modification process should be viewed as a temporary measure until organizational risk management is committed to across the entire organization. At that point, these integration points should be replaced with modifications to the processes, tools, templates, etc., to allow for effective and efficient direct integration.

This is just one way to approach this integration, and your organization may have other standards for process integration. The important thing is to recognize during this initial integration phase that the work has to occur and that there needs to be clear ownership and understanding of how it will occur.

20

Process Analysis

You are now finally ready to start the *real* work of the project—beginning the process of execution on the various project elements. However, you aren't quite ready to start writing process documents or drawing workflows. The first step is to perform the analysis of the organization in order to understand the process that actually needs to be developed. Your business case described your vision for an organizational risk management process, and that will have been a good description of the desired end state—the way the organization should operate once you have completed the work. Think of this analysis as the detailed understanding of the current state—the way that risk management is conducted within the organization today.

You wouldn't have been able to conduct the business case without considering the way that things are done now. After all, if things were already optimal, we wouldn't be undertaking this project at all. However, the business case will only have looked at the current state from a fairly high-level standpoint—an overall approach and a generalization of effectiveness. Similarly, the current situation will have been considered when the project approach was considered during initiation—the approach that we are planning to take will be based in part on what currently exists.

What we need to do now is take that analysis deeper, so we have the foundation for building the new process. This analysis will help us to understand what can be reused or reworked, what needs to be discarded, what needs to be created fresh, and what needs to be developed in terms of training, documentation, and transition plans. With an organization-wide risk management initiative, we also need to understand the differences across different departments, offices, or PMOs. Even if your project is restricted to only a single PMO or a subset of the project execution that occurs within the organization, it is still worth understanding what is happening elsewhere. It will help to identify opportunities to leverage best practices, modify existing processes and tools rather than creating from scratch, and create processes that are aligned with other areas of the organization.

Understanding the Scope

The first step in the analysis, and the step that helps us to transition from initiation to planning, is ensuring we have accurately defined and understood the scope of the work ahead of us. Some preliminary work will have been done as part of the business casing process, and that will have identified the major areas of project execution within the organization, as well as the standard approved risk management approach. The organizational integration that we looked at in the last chapter will have extended on the project proposal/business case and will have started to draw some clearer boundaries around that scope.

Now that we need to document the scope in more detail, the temptation is to assume that it is obvious. We will know what happens in the *core* group of projects and programs that form the defined portfolio (or portfolios if the organization is broken out into separate autonomous PMOs), and it is easy to assume that there is nothing else to worry about. That's unlikely to be the case. However, if we assume that the project is to develop an organizational risk management approach that will be deployed throughout all areas of the business, we need to do the following:

- Identify every area within the organization where projects are undertaken. This is the most fundamental step in the analysis. If we fail to identify any project execution group within the organization, then we immediately fail in our goal of an organization-wide process. This seems like the easiest part of defining scope. It's what we would immediately seek to understand with any background in project execution, but the danger here is that there may be small pockets of specialist functions where things have always been handled outside of the standard processes. If these have been allowed to go their own way because there is little visibility into them, or minimal impact on the larger organization, then they are easily missed; and we immediately fail in our goal of an organization-wide approach to risk

Best Practice—Analyzing the Current State

There's a reason why your project was approved—at the most fundamental level the current state of risk management within your organization isn't good enough. That can cause some interesting challenges when it comes to trying to understand the details of how risk management is being executed today. People may not be willing to admit to any inadequacies or failings in current processes and process execution, they may try to put a positive spin on things, or they may try to ignore requests for information. It's vitally important to understand what is really going on within the organization because that's the starting point for all of the work that will follow. Make the process easier for yourself and for the people involved in risk management by trying to:

- Avoid any form of comparison between processes, practices, or departments. Some will always be better than others, and there are many reasons for that, none of which are going to help you gain an accurate understanding of how risk management is being executed today.
- Remain as nonjudgmental and objective as possible. Your role isn't to criticize the way that risk management has been handled to this point—it's to improve future risk management.
- Develop partnerships with the people who execute risk management. These people will ultimately be the ones who have to work with the new processes that you develop and the sooner you are able to develop collaborative relationships with them, the more help that you will receive from them during process development. They will also be more receptive to the new processes as they are rolled out.
- Avoid assuming that variations are problems. On many occasions people develop workarounds or alternative approaches to standard processes not because of an unwillingness to follow a prescribed approach, but because the standard way of doing things doesn't work in their unique situation. Understanding the details of that uniqueness will help to ensure that the processes that you develop are able to work in that environment.

Ultimately your success depends on the ability to understand and respond to the needs of the various areas of the organization that conduct risk management and if your initial interaction with them to understand the current state doesn't start to build a collaborative relationship for the project then you are creating the potential for significant downstream problems.

management. These pockets may be in small, remote offices; they may be outlier functions from a corporate takeover; a group set up to pursue a specific opportunity; or any number of unique situations. Success depends on identifying them and incorporating them into the initiative. Even if those groups have traditionally only handled project level initiatives and not program or higher, they should still be considered here—a lack of inclusion does not mean that they are exempt from being a part of the overall corporate portfolio.

- Obtain existing risk management processes. Once we have identified each of the groups, we need to obtain the processes that they currently use. We are now starting to add depth to the scope—starting the transition into gathering the details that will help us to plan the specific activities. Although we are focused on organizational risk management with this initiative, we should aim to gain an understanding of related processes—as we saw in the previous chapter, the way that risk management integrates with the other processes will provide context for the work ahead. We should also look at project level risk processes, not just program and portfolio. In some situations, project level risk management will be all that is performed. Additionally we need to ensure that the organizational approach we develop can work with the existing project risk management. If the approach we ultimately develop for the organization is incompatible with existing project level risk management approaches, then we may end up driving project level risk management change. That is not an extension of scope; by definition, organizational or portfolio risk management includes project management. We are simply not going to change those processes if it isn't necessary.

- Obtain risk management examples. In addition to the processes themselves, we need to obtain some examples of the way that those processes are being used. For each of the independent groups, we should be looking for some examples of risk logs, risk management plans, individual risk summaries, and contingency plans. We aren't looking to second guess the work. In fact, the specifics of the risks are not particularly relevant, but we want to understand the level of detail, the frequency of review, the versions of templates that are being used, and an overall sense of the quality of execution of the process. Where processes differ, the examples will help us to understand the processes more completely. Where the processes are similar, the examples will highlight differences in execution.

- Secure details of risk management success. In many cases, this is likely to be the hardest set of data to obtain because it is frequently not actively tracked. If the organization does not have a culture of risk management, then there may not even be details of contingency or management reserves available, let alone an understanding of how the actual risk costs mapped to those reserves in recent projects. We may be forced to look at project plans for recent initiatives to try and determine the success of risk management based on the risks that triggered and the impact of those events, and that may not be easy to determine. However, anything that we can obtain here is going to be tremendously valuable—it will provide us with a baseline against which we can measure the success of our organizational risk management approach.

- Understand organizational level data. I am not expecting that the company will have an organizational risk profile at this point, but we should be looking for information on the risk environment within which the organization operates. This is unlikely to be a set of files that can be provided—rather, it will probably involve a series of interviews with executives to understand the organizational risks that they are facing, both the internal operational risks and the external risks that impact the organization. In some cases, this may be forcing executives to think about risk in ways that they haven't previously done; that's not only going to help us to understand the environment for which we are developing processes, it is also a preliminary deliverable for the project—increased organizational risk awareness.

It's pretty easy to see that this work is going to provide a lot of raw data to the project team, all of which will need to be processed, but unless we obtain it, we are not going to be able to build a full picture of the current state of risk management within the organization. What we now need to do in order to understand the scope of the work ahead of us is to turn this data into meaningful information.

I always find it easier to start with a single high-level picture and drill down to the detail I need from there. I build a matrix of the different PMOs, departments, offices, and units that have some project execution with the different elements (process, templates, examples) mapped against them. This provides an easy to understand view of any gaps, as well as allowing for the identification of a specific element within the matrix for which we can get more information. This approach also helps to understand and communicate the extent of the work ahead of us. If there are twenty different project execution areas, each with their own way of doing things, then we have much more work ahead of us than if there are just a handful of groups, most of whom have no defined process in place. Not everything fits nicely into the matrix. The organizational information usually has to be considered separately, and the quality of the data on risk management success is likely to vary considerably, making comparison rather dangerous, but the matrix is still a good starting point for the analysis. Table 20.1 shows a basic template for this matrix with a few examples—a real matrix would include links to the documents that relate to each cell in the matrix.

As you can see from Table 20.1, we have summarized the level of compliance and rigorousness with which each element is applied within each category, their similarity to any existing corporate standard (whether formal or generally accepted), and can also begin to determine whether the process, tool, or template can be used as an input for our new process. You won't be working with the restriction of needing to get your matrix to fit onto a single printed page, so you will be able to add greater levels of depth and detail and additional category columns until you have an extensive summary of the analysis that has been performed, along with links from that summary to each of the detailed elements that sits beneath it.

Once this work has been completed, we should have a solid understanding of the way risks are managed throughout the organization. We will have a sense of areas with a strong foundation of risk management already, even if only at a project level, and we will have identified some areas where the processes, and/or the way that those processes are applied, gives rise to some concern. We may also have identified some additional experts that we need to engage in the project—perhaps someone who wrote a comprehensive process for a specialist group within the organization that we didn't previously know existed. The next step is to take the analysis deeper and determine the details of the work in each aspect of process development—the scale of the work.

Understanding the Scale

In this part of the analysis, we take a different view of the information we have obtained. Instead of looking at the matrix from the perspective of each project execution area, we now look at it from the perspective of

Table 20.1 Risk management process analysis matrix

Department	Process	Templates	Examples	Support documentation	Training material	Success measures	Notes
IT PMO	Standard	Some standard, some custom	Variable quality	Standard	Standard, but no formal execution	Not tracked	Significant variation in quality of risk management, often not applied
Corporate PMO	Standard	Standard	Some missing	Incomplete	Standard	Variable tracking	General compliance with existing standards
Accounting	Standard	Standard	Consistent high quality	Standard	Standard	Always monitored	Consider using as model for rollout/deployment, excellent compliance, and execution
Product	Some standard	Some standard	Few available	None	None	Not tracked	Some standard processes in place, but not generally followed
Europe region	Custom	Custom	Consistent high quality	Custom, excellent detail	Custom, well developed program	Detailed tracking, high quality results	Still using pre-acquisition processes, but excellent compliance and results
Asia region	Ad hoc	Some custom, some missing	Variable quality and format	None	None	Variable tracking, but good results	Process is incomplete and largely ad hoc by PM, but good risk management results—need to analyze

Note—standard references corporate project risk management standard version x.x dated dd/mm/yyyy. No department undertakes program or portfolio level risk management.

each element we need to develop. We are concerned here with understanding how much of the material we have gathered from the existing processes can be reused, how much can form a foundation or starting point, and how much is going to have to be discarded. In other words, it's an analysis of how closely the current inventory of process assets matches the needs of the organization. If the concept of portfolio level processes is new to the organization, we might expect there won't be much that can be leveraged from the existing processes, tools, and templates, but that isn't necessarily the case. There may be building blocks that can be enhanced and refined. This will not only help to simplify the process development, it will provide process practitioners with something familiar, helping them to embrace the new approach.

When we look at this from a scale standpoint, we consider both the individual groups (how much needs to change for this area of the organization) and the organization as a whole (how much can be reused when we combine everything together). The following are areas we are particularly concerned with understanding in terms of what exists, and where the gaps exist:

- *The process itself.* This will be the largest piece of work and will consist of a detailed description of each element of the process, the inputs and outputs, the steps involved, and the process flow diagrams. This will be your version of the organizational risk management process as outlined in Section 2, and your process will likely have a broad resemblance to that framework. If your organization has not previously embraced the concept of a portfolio-wide process, then there will likely not be a reusable process. You should be wary of taking an entire project level process and trying to apply it at the portfolio level, as it is unlikely to be a good fit. However, you may be able to identify a number of best practices for certain areas—perhaps a particularly strong approach to contingency planning that occurs in one department.

- *The tools and templates.* These are the physical artifacts that are developed to support the process and that will be needed for every item that is produced during the execution of the process. Templates will include the *headline* items like the risk management plan template, individual risk summaries, and contingency plan templates, but there will be many others as the details are defined. Tools are artifacts designed to support the process and templates. A typical example is a checklist that is completed as the process is undertaken—a list of steps that a risk analyst has to complete or items that should be considered when developing a contingency plan, for example. We should be looking to leverage existing templates wherever possible, and we explored this a little bit in the previous chapter when we considered process integration. Practitioners of organizational risk management will likely also be engaged in project level risk management, and any similarity in the tools and templates used will help with the understanding. However, this can't be at the cost of sacrificing needed functionality at the portfolio and program levels—existing templates may form a foundation but will likely need some modification. Examples of some common tools and templates can be found in the Web Added Value™ Download Resource Center for this book that is available at www.jrosspub.com.

- *The training and educational material.* This is the material that will be used to train people on the process. Obviously, if the process changes, then this material also has to change. However, we may find that there are imaginative approaches to training delivery that can be reused or exercises and examples that can be adapted to the new process. This section will also include supporting material—guidelines, hints and tips, and frequently asked questions (FAQ) documents.

At this early stage of the project, we may not fully know what can be reworked or reused. We will need to revisit this work as the project continues, but we do need to go through this process to eliminate some of the items we know will be inadequate and identify potential candidates for further analysis as the project continues. That will help us to keep the information that we are working with at a more manageable level and will help us to focus on the areas of the organization likely to offer us the best starting point for our new process. We shouldn't assume that we are modifying what already exists, but neither should we ignore what is already working.

Once this work has been completed, I will again go back to the matrix that we looked at earlier, and I will add this analysis to it. At this point, I use color coding to highlight areas we know have to be replaced, the areas we are fairly sure can be adapted or reused, and the areas we aren't yet sure about. A simple traffic light model works for me with each cell in the matrix identified based on the strength of the current element, but feel free to get as complex as you like, as long as you can still readily interpret what the matrix is telling you.

I also conduct a *sanity check* at this point to make sure the decisions we are reflecting on the matrix are logical. If we have determined that the process that is being conducted by one particular area of the business is unusable and must be replaced, yet we think that the tools and templates used by that group are a solid foundation for our new templates, we may need to take a second look at that area because those appear to be inconsistent data points. It seems unlikely that a process that we consider to be unsuitable for use at an organization-wide level would have templates we think are a solid foundation for our new approach.

Validating the Approach

Once we understand the scope and the scale, we should take the opportunity to validate, and if necessary adjust, the planned approach that we developed during the project initiation. The work of a project is never completely linear—we need to go back and consider whether what we have learned here has changed one or more elements that we developed during initiation. As we have completed our analysis of the processes, tools, and templates that are currently being used in the organization, we may well have identified new stakeholders and new areas of the organization that need to be included in the scope. This will require us to review and potentially update the communication strategy and the way that we are going to integrate our work with existing related processes. It may also identify a number of different resources that we want to try and engage in the project.

21

Process Development

We are now ready to start the work that will form the largest part of the project—the development of new and/or revised processes. At the beginning of the last chapter, I described the business case as providing the vision for an effective organizational risk management approach, and the process analysis as the examination of the current state of those processes. It follows that the processes we develop need to allow the organization to bridge the gaps between current state and desired end state.

This chapter more than any other can be considered to be a generic guide to the development of an organizational process, and in some ways to a process development initiative at any level of the organization. The examples relate to risk management, assuming that this will be your primary focus; however, you should be looking for ways to create an approach to developing and implementing organizational process that can be used for any organizational process deployment, or indeed for an entire organizational methodology. That leaves you free to focus on the key issues of the process that is being developed rather than on the way that it is being developed.

The matrix described in the previous chapter is a useful summary of those gaps and can form a checklist of the work we have performed after we develop the new processes. However, we should be careful not to review the matrix as a listing of the work performed. If the focus becomes to check off items in the matrix, then we will develop a series of tactical steps that may be effective but will not necessarily form a cohesive organizational risk management approach that is focused on improving the quality of strategic risk management. Instead, we should ensure the overall approach addresses each of the gaps outlined in the matrix, not address each of the gaps one by one to build an overall approach. Put another way, we need a strategic risk management solution that, when applied properly, solves the individual tactical elements as part of the overall solution.

You should already be clear on what the framework for your organizational risk management approach should look like—there is a detailed walkthrough of the process elements and the different considerations at the portfolio, program, and PMO levels in Section 2 of this book. However, you will likely have to adjust that framework to the unique complexities of your own project execution environment, and you will certainly need to add detailed steps to allow for the integration with the rest of your project execution methodology. Consider the framework outlined in Section 2 a skeleton—the common structure we all share and on which the individual elements that make everyone (and every process) unique are built. That's what this chapter addresses, however, you still aren't ready to start writing process—you need to determine the process framework first.

Defining the Process Structure Framework

The process structure framework defines the way the processes will be developed and implemented. Process development is more than simply process flow diagrams and templates. The process needs to be created and rolled out to the organization in such a way that it can be understood and accepted within the organization. That means we need to create a structure that works for the people who will ultimately be executing the processes that are developed.

In some ways, this is an extension of the portfolio management maturity concepts we looked at in Chapter 18, although there is no direct connection. An organization that has an immature project execution approach is extremely unlikely to have a robust process structure, and you can assume you will need to develop a new structure that can support organizational risk management and future process development work. However, even organizations that have a relatively high level of project execution maturity may not have any consistent or reliable process structure, instead using as many different template formats, process flow techniques, and reporting standards as there are processes.

In determining the framework for the process structure that you need to develop, consider the following:

- Existing process structure within the organization. If there is already a common approach to processes within the organization, you may find it beneficial to use that. If the people who need to use the organizational risk management process recognize the templates and the diagramming look and feel then it will immediately provide them with something familiar with which to connect. That in turn will help them to feel more comfortable with what's happening, if only because it's not completely new. On the other hand, you don't want to force yourself to use a process structure that was designed for project level processes, is out of date, or simply doesn't suit the processes that you are looking to develop.

- Areas of the organization that will be consuming the new process. Even something as simple as using a template format that is only used by one particular PMO can unintentionally cause difficulties with other areas of the business who might feel the new processes were designed for another group and are now being forced on them, even though they weren't the key focus. On the other hand, if a large percentage of the organization is familiar with one particular process structure, then you can make implementation and acceptance easier by using that style.

- Relative complexity of the processes you are developing compared with existing processes. If this is the first organizational level process that the organization is implementing, you may need to develop a new structure because the existing approaches will simply not be robust enough. At the same time, if you can use the same process structure foundations, then it will help to make the new organizational processes feel as though they are an extension of the processes that are already in place.

- Process elements you are developing. Your structure needs to allow for all of the items you are developing—the process itself, the support material, and the tools and templates. Creating a common format or branding will help to establish all of the items as related to one another and will start to create a look and feel to the organizational risk management process that people will recognize.

- Management and maintenance of the process. Presumably, you will be handing ownership of the process off to a PMO or the organization's EPMO to manage once the project has been completed, and that group will need to be able to manage the framework and structure effectively and efficiently. Your structure should make it easy for audit and control to occur, for process enhancement to happen, and for reporting and tracking to be managed.

While there is a lot of appeal to trying to use a structure that the organization already uses, there are also a lot of challenges to that approach. One of the major problems is that organizations simply don't invest a lot of time and effort into a process structure, so what you observe as existing templates may simply be the result of one person drawing all of the process flows and no one ever updating them. You also have to consider that just because a template is in use, that doesn't mean that the organization likes the approach. For the same reason as above, organizations don't invest in process structure, and the likelihood is that no one spoke to process practitioners about how they would like to see the processes documented, the templates set out, and so on.

We discussed in Chapter 19 the importance of the stakeholders and here is a real opportunity to reach out to a subset of the passive stakeholders to gain their input on how the process artifacts should be structured. This approach will help to ensure the process structure is better aligned with the needs of the consumers of the process, and it helps to engage the stakeholders in the process—allowing them to recognize that their needs are important. Just like in any other project, you shouldn't hand control of the process over to the stakeholders, but they can provide you with valuable input that can be used by the team in determining the process structure.

Your process structure will vary depending on your organization's needs and expectations, but these are the items that I consider to collectively offer a comprehensive process structure:

- *Process template.* This is the written format for each process document. This is the key detail document, and it's important to have a consistent template so that users can quickly navigate the document for the information they need. Remember that this will be a reference document most of the time, so it needs to be clearly laid out into sections that make it easy for users to find the information they need. Typical sections may include purpose (what this process is designed to achieve), inputs (items required to complete this process), outputs (items generated or changed by this process), process steps (the sequence of steps that should be followed), and templates (the related items that are used during the execution of the process).
- *Process naming and coding convention.* Even if you are developing the first formal process that the organization has ever had, there should be an expectation that you will eventually develop a complete methodology. You should create a standard

convention for processes and process elements that make it clear which process each element belongs to and how all of the different elements are sequenced and roll up to the higher level elements. For example, organizational risk management (ORM) may consist of five elements (ORM1 through ORM5), and the first element may have three steps (ORM1.1 leads to ORM1.2 and ORM1.3. ORM1.3 then leads to the second element ORM2.1).

- *Process flow diagram standards.* Process flows are the graphical summaries of the process—the figures in Chapters 8 through 12, for example. These should be consistently drawn using the same approach and format and should clearly identify how they relate to one another. There should be high-level summary flows as well as more detailed flows of detailed process elements.

- *Template standards.* It's difficult to insist that all templates have to comply with the same look and feel because they serve different purposes—a template for a risk management plan is not going to look like the template for a detailed risk summary, even if the fields are the same. However, you can create standards for key elements of the templates—naming conventions, terms and acronyms, and version control, for example. Even something as simple as a filename that makes the template's purpose clear and what the latest version is will go a long way toward making the process easier to accept. One note on that point: a lot of people use a date to determine the latest version, but that doesn't work (how do you know there is not another version with a newer date?). Instead, use something like *current* for the latest version and archive retired versions by date.

- *Training and support material format.* People will find it a lot easier to learn about new processes if the training is provided in a consistent manner for all processes and all process elements. If the method and approach to training varies significantly, then people will need to adjust to these changes and that takes time, as well as detracting from where their focus should be—learning how to use the new processes.

- *Storage and access locations.* One of the problems with maintaining standards for processes is the fact that documents and templates end up being scattered across the organization, and we soon end up with multiple variations in use in different areas. You need to create a central location that is logically structured and available to all employees where process documentation and support material can be accessed by anyone who needs it. That means the location needs to be somewhere all staff have access to regardless of their office or division, and it needs to be somewhere easy to find (if on a shared network folder or intranet page, then add the mapping/favorite link to part of all users' technology profile).

The creation of standards can be a difficult exercise, especially if the organization already has a nominal approach that is used but that we consider inadequate for our process. You will have PMO/EPMO resources on the project and in your stakeholder group who can agree to a change in the standards, but we don't want to expand the scope of our project to develop standards for the entire organization. There will need to be agreement to a future initiative to bring the standards in line (or add them to other processes if they are absent).

If you are defining a new standard that will ultimately be deployed across the entire portfolio execution environment, make sure you consider the fact that eventually the standard approach will need to extend beyond organizational risk management and encompass other process areas. The implication of this is that there needs to be a way for organizational risk management to stand out within this common standard—an easy way to identify a process or template as relating to the organizational risk process. This isn't difficult—a colored border or a prefix to every document name will easily achieve it, but it's also easily forgotten with a focus on a single process area and harder to retrofit later on.

Process Creation Basics

You are finally ready to start creating your processes. By this point, we will have an understanding of the scope and scale of the work ahead of us, we will have a reasonable understanding of how the organizational risk process needs to interact and integrate with other process areas within the organization, and we will have a team of resources and an understanding of their skills and experiences. We will have identified the stakeholders with whom we need to interact, and we will have the framework for the processes and support material we will be developing.

We now need to decide how best to utilize the team that we have to create the best possible processes within the constraints of the project. There are a few basic tenets that will help to shape that determination:

- Process definition is not a solitary exercise. Sure, one person could sit down and write out the processes; in fact, if you are going to do that, you could just copy what I have in Section 2 and add a little detail for your organization. However, the best process will come from discussion and debate, from proposing solutions and then challenging them to find the flaws.
- Processes cannot be defined in isolation. By definition, a process is a sequence of a number of different activities that flow together to create an integrated sequence. You can split your project team into a number of different workgroups in order to try to develop process elements in parallel, but those workgroups will need to frequently work together to ensure that they are aligned and that the processes being developed flow together smoothly and effectively.
- You won't get it right the first time. No matter how much work you put into the processes that are being developed, you will need to make revisions once you start testing them. By planning for frequent *sanity checks* within the team and with stakeholders, you can make the adjustments as you go and minimize the extent of the changes that are required when the end-to-end process is ready for testing.
- You will be forced to compromise. The processes you are developing can never be perfect. You are developing a process that needs to operate within an imperfect organizational environment, not in an artificial environment where anything is possible. That will mean there will always be a trade-off between the *perfect* process and the most efficient process. Ultimately, you are looking for the most effective process—the point where you are getting the best risk management that you can for a realistic cost of execution.
- You can't please everyone. This isn't exactly news to anyone who has managed a project before, but when you are building processes for multiple areas within the organization, you will always have some elements who feel that their needs are not fully met by the processes that are built. Even when you are only developing processes for a single PMO, there will be individuals who feel that their needs and/or recommendations are being ignored.

Don't Reinvent the Wheel

Section 2 of this book outlined an organizational risk management process for you. That process is pretty complete, but it is also pretty generic. It is geared around a conceptual organization that simply doesn't exist—your organization is not cookie cutter, it is an individual entity that is constantly shifting and evolving and needs an organizational risk management approach that reflects those unique elements. However, at the same time, you have a lot of similarities with that generic organization. You have multiple projects with conflicting priorities and demands, you exist within an environment that imposes a number of risks upon the organization, and you only have a certain tolerance for all of the risks. You have limited resources to commit to project and you have limited resources available for risk management.

As a result, you can view the process that I outlined as a solid framework that you can enhance and adjust until it is well suited to your needs. That process may not be something you can simply copy and paste into your templates to create a living, breathing organizational risk management process, but it is a foundation that provides you with the basics and allows you to focus on the details and application rather than the base structure.

Obviously, I think that this process is a good approach because I wrote it, but I'm not going to be offended if you want to make some changes to it. In fact, I encourage you to make modifications so that it becomes better suited to the unique challenges faced by your organization. If you wish, discard this process entirely and substitute a process that you already have within your organization, or that you have been exposed to in the past. Just try to avoid creating a new process from a blank piece of paper if you possibly can. You will end up spending a huge amount of time and effort getting to the point presented to you earlier in the book, and you will likely find that there is not that much tangible difference in the approach you develop.

Even if you have decided to use an approach to process development that involves defining all of the details as you go, you will be well served to select and adjust your process framework before you start detailed process development work. You need to understand the fundamental direction and objectives of the process before you start defining details.

From Framework to Process

Once you have your process framework in place, you can start adding the details. To continue my analogy from earlier in this chapter, if the process framework is the skeleton, then these details are the features and characteristics that turn that skeleton into something unique and individual.

Best Practice—Effective and Efficient Process Development

Your project may be the first organizational process that has ever been attempted within the organization. It may even be the first formal risk management process that has been implemented, but that doesn't mean that you have to start from a blank piece of paper. Improve your chances of success and help your team focus on areas where they can truly add value by adopting as many of the following as possible:

- Align your process structure (e.g., style, template format, naming convention) with that which the organization and its employees are already familiar.
- Leverage existing processes and process elements where possible—from within your organization, from previous experiences, from the framework in section 2 of this book, etc.
- Ensure that processes are always developed from the perspective of the practitioner—they are being implemented to improve the chances of organizational success, but that can only occur through the actions of the people responsible for the execution of the work.
- Focus on the support infrastructure for the process—the processes themselves are not the only deliverables, and success depends just as much on the training material, the alignment with other processes, and the buy-in of stakeholders as it does on the right risk identification, analysis, management, and contingency.
- Involve passive stakeholders in your work and look to gain their input and support early and often.

An organizational process will always be disruptive to the way that work is undertaken, and a process that is applied as frequently and as widely as risk management can be especially impactful. By planning from the outset to minimize the disruption and maximize benefits, you can go a long way toward being successful before the first process flow is ever drawn.

Remember the basic tenets that we discussed. This work shouldn't involve a group of people running away to draw diagrams and build templates. Instead, this work will break the framework down into a number of individual elements based on the approach you have chosen to take—think of an element as one of the boxes in the process flow diagrams in Figures 8.1, 9.1, 10.1, 11.1, and 12.1, although you may obviously have made adjustments. For each of those elements, the process we develop should answer the following questions:

- What should the implementation of this element within our organization look like?
- Who should be responsible for the execution of this element?
- What information does this element need to be executed successfully?
- What new or modified information does this element generate?
- What tools and templates are needed to execute this element effectively?
- What are the exceptions to this process?
- What supporting material and processes are required to assist people in executing the process?

Once we have answered each of those questions, then we are well on our way to creating a fully developed process. Let's look at each of those questions in more detail.

What Should the Implementation Look Like?

This is the biggest question, and it is where we have to start the process of adding the features to our process framework. In the framework, this is simply one box in the process flow diagram, but in reality you will probably need to add more details to that—a lower level that may add more steps and will definitely add additional parameters to address the specific situations you have in your organization.

Consider, as an example, the first box in Figure 8.1 (risk identification), which is the individual review—the process of having someone individually review the portfolio or elements of the portfolio and identify where previously unidentified potential risks exist. It's likely you will have a similar process element in your framework even if you chose not to use my process as your starting point. You need to have a way to identify new risks and get them into the risk management process, but you need to define more detail than we currently have if the process is going to be usable by your portfolio execution resources. This box was about asking yourself *"based on my understanding of the portfolio, my study of the factors that affect the portfolio and programs that form the inputs to this process, and my skills and knowledge in my role, what risks exist that have not yet been captured in the risk list?"* That's fair enough, but you may need to break that box down into a number of steps to better define the review process:

> **Best Practice—Don't Forget . . .**
>
> This chapter covers a lot of elements of detailed process development, but remember that this project is all about improving the quality of risk management within your organization. Throughout the project you should be continuously asking yourself:
>
> - Are we focusing on the areas where we will make the most difference—strategic improvements to areas of traditionally high risk exposure?
> - Are we implementing processes that can be adopted and integrated with the organization's approach to project execution?
> - Are we working to ensure that our stakeholders genuinely feel as though they have a voice and that their needs and preferences are being considered?
> - Are we creating a foundation that can be built in the form of either a more widespread implementation of risk management or a more comprehensive strategic portfolio execution approach (or both)?
>
> And, of course . . .
>
> - Are we managing the initiative like a well-executed strategic project (with particular regard to effective risk management)?
>
> The project manager for this initiative should be asking him or herself that question frequently. It's easy to take process development initiatives down *into the weeds* and spend inordinate amounts of time discussing template layouts and process diagram design. While those items need to be considered, they are unlikely to materially impact the quality of the process in a way that can't be refined through continuous improvement. However, you only get one chance to get the process right before your credibility is questioned!

- How to ensure that all of the potential sources of risk are identified
- What you should be looking for as a sign that a risk might exist
- How best to document the potential risks that are identified
- How to conduct a preliminary estimation of related risks and relative importance of the risks

In addition to the various steps, you may also need to provide the practitioners executing the process some parameters around the work that needs to be undertaken for this process element—additional characteristics that add to the overall process. As an example, unless you provide guidelines, some people will spend just a few minutes on the work while others will spend days going over project documents with a fine-tooth comb. Neither of those approaches is appropriate, but unless you provide some guidance, people won't necessarily know that. This shouldn't be a hard and fast rule. You don't want people to think they have to keep a stop watch on their actions, and everyone works at a different speed. You also have to consider that rules tend to create a negative connotation that leads to resentment and a sense of being micromanaged. Instead, having a guideline that provides a suggested range of effort and/or elapsed time for each step helps to provide some context without applying unnecessary rigor.

Other parameters applying to this process element might be the typical number of organizational risks that you might expect to see (this should be a broad range because there are so many variables, but it will help people understand if they should be looking for 1, 10, 100, or 1,000) or the level of detail that people should be capturing during this process element. A good test to identify parameters that might apply to a process element is to consider the questions people might have when they are asked to perform the work. Any question that involves some kind of quantification is a likely candidate for a parameter or guideline to be defined.

Who Should be Responsible?

When you are creating any process, you need to consider who needs to be involved in each process when it is executed. I have written the question as who is *responsible*, but more accurately, you need to consider both responsibility and accountability. Those terms are often used interchangeably, but they are different:

- The person responsible does the work.
- The person accountable ensures that the work gets done.

In order to ensure that you have an effective process, you need to ensure there is a single person or role accountable for each piece of work. The buck has to stop somewhere, and that somewhere has to be readily identifiable—hence, one, and only one, person accountable. On the other hand, there may be multiple people responsible for the work. People will need to collaborate to get the work done and each of those people has to contribute to the execution of the process. You will also have individuals who are associated with the work but who don't have an ownership stake; we'll look at those shortly.

We need to be as specific as possible without being restrictive. I have seen a lot of processes that assign an element to *team member*, *PMO*, or (my personal favorite) *expert*. This is neither helpful, nor accurate. There are few elements that can be performed by any team member regardless of skills, and if we assign a task to an entire group, no one will end up doing the work. The problem with assigning a task to an expert is that it doesn't define what makes a person an expert. At the same time, we need to be flexible enough to ensure we don't end up limiting the risk assessment work, which is likely to be a sizeable amount of effort, to only a single person by having a definition of the person responsible that can only be met by one person in the company.

You may not need to specify owners at the lowest level of the process. There may be a series of steps that all require the same skills and experience, in which case it's perfectly acceptable to assign responsibility and accountability at the point where those steps accumulate. At the same time, you should avoid defining multiple owners at a rolled up process element level if in reality they each own separate steps in the process.

RACI or RASCI charts helps to make this clearer and defines all of the roles. RACI stands for:

R—Responsible
A—Accountable
C—Consulted (sometimes Contributes)
I—Informed

When we add the S for RASCI, we are adding an additional role that is only required for some elements of the risk management process:

S—Sign-off

The chart that we create is a matrix between the process elements and the roles that are engaged in those process elements. Table 21.1 shows an example of a RASCI chart for the high-level elements of risk analysis from Figure 9.1. This is intended simply to be indicative; your chart will likely be more detailed with the left-hand column having a more detailed breakdown of the process elements and the top row having additional roles. I also try to define each of the roles to avoid the potential problem described above of having a definition that is too generic—for the same reason, you may need to consider qualifiers like junior/intermediate/senior or similar.

There are a few more guidelines that will help you ensure that your assignment of responsibility and accountability goes smoothly:

- Provide guidance on the necessary skills. By providing a framework description when we develop the process of the needed skills to execute that process, we can help to identify the right

Table 21.1   Sample RASCI chart—high-level risk analysis

	Roles / Responsibilities					
	Risk analyst	Subject matter expert	Portfolio manager	PMO representative	Program manager	Project manager
Perform qualitative analysis	R	R	C	A	C	C
Perform quantitative analysis	R	R	C	A	C	C
Prioritization of risks	R	C	A	C, I	C, I	C, I
Determine management approach	R	C	A	C	C	
Review & approve	I	I	A	R		

RASCI Chart Rules
- Every task should have a role that is identified as accountable, and only one
- Every task should have a role that is identified as responsible, and there may be more than one role responsible
- The same person cannot be both accountable and responsible

person for the task. For example, we may recommend that risk prioritization be assigned to a "team member with experience analyzing risks, preferably also someone familiar with determining impact and recommending the appropriate management approach." This establishes a minimum standard (risk management experience) along with some additional recommended experience, but it still remains flexible enough for the individual managing the execution of the process to be able to apply their judgment in the specific assignments.

- Specify functions. Sometimes you will need a mix of skills on the same element of the process. For example, the analysis of a risk will require someone who understands the project, program, or portfolio elements of the risk, and one or more people who are subject matter experts in the area of potential impact for the specific risk. In this case, you should recommend the role that each person has on the process element to help the practitioners understand their roles.
- Build logic into the assignments. In the organizational risk analysis process (Figure 9.1), we built in a formal approval step that confirms the analysis, prioritization, and management

approach. This approval process exists for a reason—to make sure the work is accurate and complete before we move on to the management phase. The assignment of resources to this approval process needs to reflect the work that is expected here. The approval doesn't necessarily have to be done by a *senior resource*, but there is likely a minimum skill and experience level. There should be a check in place to ensure there is no opportunity for someone doing the analysis, prioritization, or determining the approach to approve their own work. This should be made clear in the definition of responsibility.

- Accountability should always be to a single specific role. While flexibility can be used in determining responsibility for a task, accountability should be more restrictive. The process should make it clear that accountability needs to be with an individual rather than a group, but that should be an appropriate owner—someone close enough to the work to be seen as the most directly accountable person. I have seen processes where the head of the PMO or the portfolio manager was made accountable for everything, and that's not realistic because they aren't going to be that involved with all elements of the work. Of course, they have a level of accountability, but if we aren't going to identify the first person in the accountability chain, then why stop with the portfolio manager—why not make the CEO accountable, or the chairman of the board? Most of the assignments of responsibility will be recommendations. You will be suggesting the skills and experience that are best suited to the execution of each process element. The final decision about whether to follow those guidelines or assign a different role or skill set will lie with the person managing the specific execution of the process and who is, therefore, closest to the unique situation that needs to be addressed. However, there are two resource related items that should be considered hard and fast rules with no ability to override them. I would include these rules in any compliance audit:

  - Every process element has an individual (and only one individual) identified as accountable. It is vital to know who to go to in order to resolve any issues with the execution of a process, especially when that process has the potential to impact the success of the entire portfolio. We cannot get into a situation where stakeholders or the portfolio manager are running around trying to find the person with the authority to make a decision—it simply must be clearly defined up front.

- Self-approval should never be allowed. Peer review and approval is perfectly acceptable within a process, but there should never be a situation where someone is authorized to approve their own work.

As a test of the success of your proposed resource assignment recommendations to a newly developed process, you could review with a group of stakeholders from the program and portfolio levels and see whether they can identify appropriate resources for each of the process elements from their current initiatives. If they are struggling to identify people for some of the process elements, then your guidelines may be too restrictive. If they are coming up with a long list of candidates, then the guidelines may be too generic.

What Information Is Needed?

Once you have determined how the steps within each element will be executed and have identified the owners of those steps, you need to think about what information those people require in order to complete the work. This is an area of process development that can be difficult—in the somewhat artificial world of defining the process, it is easy to misjudge the information that is going to be required. This may result in us identifying a huge number of different items that will cause the process to bog down, or it may lead us to assume that the quality of available information will be so good that only a small number of items are required. Neither is likely to be accurate.

With risk management, this can be even more complex because the unique elements of the risk can require different types of information to be consumed—item-specific information in addition to process-specific information. It can be tempting to generically group this into a broad category of *other risk specific information*. That's a perfectly acceptable approach as long as the risk process provides examples of what kind of information might fall into this category—it's our job to make that clear during the process definition, not the job of practitioners to figure it out when they are executing risk management in the midst of a complex portfolio. Information is the raw material without which the process step cannot be successfully completed, and this will ultimately become the required inputs to each step in the process. For now, let's just think of it as information.

At the most fundamental level, the execution of a step in the process is about applying skills and experience to one or more pieces of information in order to generate new or modified information that can then move to the next step in the process. By applying the right skills and experience at the right time, in the right way, and in the right sequence, we deliver effective risk management. Without any one of those elements, we have no chance to succeed. Yet so many processes focus on the work that happens in each step and virtually ignore the information that is the required *fuel* for the process.

For each step that we define in the process, we need to consider the following:

- What information is going to be transformed or consumed by this process? This will be key information that is absolutely mandatory for the process to be successful, and it should be fairly easy to identify—implementation of contingency requires a contingency plan, updating a risk management plan requires a risk management plan and updated information, and so on.
- What information supports this process? This is much harder to determine because the information in this category is not physically consumed or transformed by the process. Rather, this is decision support information—material that will help us to execute the process step efficiently and effectively. This is where the risk specific information often factors in—think about how variable the decision support information for risk analysis might be.

Let's focus on that second category because that is where we are going to have to spend most of our time in determining the required information for each process step. There will be a lot of information potentially available to us, and it can be tempting to identify all of it as required by each process step. However, there are a number of costs associated with taking that approach.

First, we have to consider that every piece of information attached to a process element will need to be reviewed and considered by the people who end up executing the process. That will take time and effort and result in a process that costs more and takes longer. That's not necessarily a bad thing, but we have to ensure that the time and effort spent is adding appropriate value to the overall quality of our organizational risk management.

Second, we may be faced with different pieces of information that appear to provide conflicting advice, are of different levels of quality (although not always obviously so), or may be only peripherally relevant. A perfect nonwork related example is an Internet search. Search for something subjective like *best restaurant* in your town and read the first link that is returned. You will then have an idea of what the best restaurant is. Read two links, and you will probably have less of an idea because there will be conflicting advice. Read a whole page of links, and you will likely wish that you had never started looking in the first place! The situation is no different in the work place, so we need to be careful about referring to too many pieces of information.

Third, we have to consider the availability of information. Unless we have current and accurate information available to us, then we will either be putting the process on hold until the information is available, or we will be using inaccurate, incomplete, and/or outdated information—none of which is a good thing.

That doesn't mean that we should avoid identifying support information. In fact, it is impossible to execute organizational risk management without considering this category of information: the organizational risk profile, the organizational constraints hierarchy, and the database of historic project information. What it does mean, however, is that we have to be careful in defining this category of information—making a distinction between information that is required and information that is recommended, and providing practitioners of the process with the flexibility to apply their own judgment in the execution of the process.

In determining the mandatory decision support information, we should focus on information that is core to the successful execution of the process. In most cases, this will be fairly high-level information and may well have been identified in the inputs column of each of the tables in Chapters 8 through 11—things like the plans, goals and objectives, resource allocations, and the risk profile and constraints hierarchy mentioned previously in this chapter. Don't rely solely on that list. Your organization may have additional items that fit in this category, but in determining your required decision support information, you should focus on items without which the chances of success for the process are significantly reduced.

For the optional decision support material, you should still aim to provide process practitioners with guidelines to help them determine which items are the most important for them. Practitioners will have the flexibility to decide for themselves which are important in their specific scenarios, but you can provide assistance through the categorization of different *quality* levels. This is somewhat subjective, but there are some broad assumptions that you can make.

For example, we all know that organizations should retain a record of past projects such as key documents and lessons learned. Most organizations will retain something, but there are huge variations in the quality and consistency of that data. Previous projects can be extremely helpful when it comes to organizational risk management because they will provide information on the analysis of previous similar risks, risk management approaches that have been attempted along with the success or otherwise of the approach, the impact of triggered risks, and the symptoms that characterized a risk about to trigger.

Clearly any one of these pieces of information can provide valuable decision support information in a number of different elements of the organizational risk management process, but only if the information is readily available. If previous project information is spread about in multiple different locations without any easy way to retrieve it, then it likely won't be considered a highly recommended source no matter how useful the information could be—the chance of being able to find relevant and helpful information is just too remote. On the other hand, if all information on archived initiatives is stored in a central, readily accessible electronic database indexed and searchable by multiple characteristics, then the previous project information should be a high-quality, highly recommended optional source—perhaps even a mandatory source. Only you know which situation better reflects your organization. If your organization doesn't have a structured central repository of archived project information then you are significantly increasing the risks to all of your initiatives because you are making it harder to learn from experience and prevent mistakes from being repeated.

Once we have identified our complete list of mandatory and optional information items, we need to ensure they are documented as inputs at the right point in the process. All of these items need to become inputs at some point—that's how the information becomes available to the process. Identifying the right point is key—too early and it creates *noise* and potentially loses the value that it should add to the process; too late and that value isn't there when it's needed.

Of course, an input isn't restricted to just one step in the process. It can be introduced at one step and then be carried through several subsequent steps, contributing value at each point. This is different from inputs that are transformed or consumed within the process. These are still reference elements, but they assist multiple steps, potentially with value being added at each step, if only through the identification of how the input can assist in the process execution.

Establishing where the information should first become an input to the process should be based on an assessment of where the decision support capability that the information provides delivers a tangible benefit to the step of the process that is being executed. This may be somewhat subjective at times, so this type of input should always be a simple guideline, with the practitioners having the freedom to consider the information earlier or later as they feel is appropriate. Similarly, when the value of the information diminishes, we should recommend removing it from being an input to that process step with the people executing the process having the ability to modify that decision.

We have focused in this section on the information type inputs rather than the more direct inputs that actually have the process steps applied to them simply because those are more straightforward to recognize and understand. However, they still need to be included in our overall process as inputs, and we need to ensure they are included at the right point. There is also the possibility that these items must reach a certain standard or level of completion before they can be used by the process, even if that means that the process has to be delayed until that point is achieved. For example, consider the individual risk summaries generated in risk identification that become inputs to the risk analysis process. Unless those summaries have a sufficient level of detail and completion, risk analysis can never be successful, because there is simply not enough information available. As a result, your process should include some kind of gate that validates that each item meets the standard to pass through as an input to the next step in the process. We are going to look at this again later in this chapter.

One final point on process inputs. We have considered here the categories of items that are subjected to being changed by the process and items that simply support the execution of the process, but we should be careful not to treat these as exclusive groups. An item that performs a decision support role in one part of the process may be modified by another step in the process. The organizational risk profile is an example. It provides us with an understanding of the risk situation within the organization (e.g., risk capacity, reserves), but it will also be changed as a result of the analysis, management, and triggering of risks.

What Information Is Generated?

It follows that after we have considered the information that acts as inputs to each step of the process, we have to consider the information that is generated within each element of the process—the information that will ultimately become the process outputs. We can't treat inputs and outputs as completely independent because an output from one process element may become an input to one or more later elements (or an input to another process through the process integration concept that we considered earlier in this section). However, there are differences we have to consider from a process development standpoint before we tie inputs and outputs together.

We have more control over this category of information than with the inputs because by definition we are doing something within the process element that affects it. In some cases, we may be creating the information from scratch—a risk summary during risk identification or a contingency plan in the contingency planning step, for example. In other situations, we will be modifying an existing piece of information—updating a risk management plan or risk list at multiple points in the process, for instance. However, the control that we have is not absolute. If we are transforming an existing piece of information, then we are reliant on the quality of the existing information (the input). As we discussed in the previous section, the use of a gating process can help to ensure that pieces of information only become inputs when they reach a minimum standard, and by implication that may mean an output is stopped from leaving a process until more work is done on it to improve the quality. This work should always be done in the right step. It's not the responsibility of a downstream process step to try and correct lower quality work from earlier process steps, or from other processes. If we are implementing organizational risk management as a first step in a new standard of organization-wide processes, then we may well find that existing processes that integrate with portfolio-level risk management are inadequate and need to be identified as requiring upgrade in a later phase of the methodology development. However, we should not be building those enhancements into a risk management approach.

This leads to one of the key elements of the process development work around the information that is generated by the process. While the steps in each element of the process are focused on what actually needs to occur, the definition of the information that is affected by that work has to focus not simply on what is created or modified, but rather on what that information should look like after the process step has been completed. That's a subtle difference, but an important one, and an example may help to make the distinction clearer.

Suppose we are looking at the risk update process within risk management, a process that is triggered if we note during risk management a need to change the risk management approach or if we observe a change in a risk's environment. It's logical to assume this work will involve an update to at least the risk management plan and the individual risk summary for the risk that has changed. The process steps should provide the framework for those changes, but it should not simply specify what needs to be done—changes to relevant fields in the risk management plan and the risk summary. Instead, it should define the end state for those pieces of information—the quality expectation for the plan and summary after the change has been made.

This expectation should not be defined in terms of specific content of the documents themselves, but in terms of ensuring they are fit for the purpose they are designed to serve, together with some indication of what that means. In other words, the process focuses on delivering outputs that meet a standard, not on the mechanics of changing or updating the information that becomes an output. This puts the focus where it should be, ensuring that the outputs of a process step are able to contribute effectively downstream in the process. In this case, that the risk management plan and individual risk summary provide downstream process elements with a clear understanding of what has changed, how it changed, why it changed, what the related impacts were, and what other work may be necessary to be triggered as a result.

It's important to get this right during the process definition work. When practitioners are following the process, they will inevitably focus on their individual work elements, an immediate, or at best short-term, focus. Unless the process is defined in such a way that the people executing the process are focused on the downstream impacts of their work, we will not end with the best possible process.

In the previous section, we talked about decision support information being an input, but because it is not transformed or created during the process steps, it should not be considered as an output of the process. It just remains within the steps of the process where it helped with the execution. If the information is needed again in a downstream step, then it will be considered an input to that step as well. However, in some scenarios, we may be adding new items to the support information, and here things can get more confusing.

An example would be the information we obtain from the project archive—historic information on different elements of previous portfolios, programs, and projects. Throughout the organizational risk management process, we will be creating new entries for the archive such as additional data points on the success or failure of risk management and additional track records for vendors. We have a situation where we are not changing the support information during the process, but we are creating new elements of support information. Shouldn't that be considered an output to this process and therefore an input to future steps?

The answer is yes, and no. In situations like this, we should consider the additional information for the project archive an output of the step that we have just completed in risk management and an input to the totally separate processes that should be in place for updating and maintaining the project archive. The relevant support information for subsequent steps should still come from the project archive as the *official* source, which in turn should include updates from all sources (not just this organizational risk management process). Clearly, there may be some delay to the updates, but an effective and efficient organizational approach to process execution should minimize this.

What Tools and Templates Are Needed?

Many process development initiatives leave the development of tools and templates to the end—finalizing all of the process elements first and then worrying about the documents that capture the information and facilitate the work. If that approach works for you, then go right ahead. I personally find it easier to develop the tools and templates in conjunction with the processes. They can't be finalized until after the process work is finished, but it seems that first draft documentation is such a natural fit with the work that we have just completed in the two previous sections. While the inputs and outputs are fresh in our minds, it would seem to make sense to try and capture the framework for the transformation that occurs during the process. Let's start by making sure that we are clear on what we are talking about when we refer to tools and templates.

Many people will use tool and template interchangeably, but they are actually distinct items. A template is a document structure that supports the process step or steps with which it is aligned. Whenever a process requires information to be created, modified, or updated, a template should exist to capture that information. We are all familiar with form type templates that provide predefined fields that the person executing the process needs to complete. The template should also include guidelines for how to complete the template to ensure the data entered is complete and sufficiently detailed. The key for me is that once the template is completed, it becomes an output to the process. That's one of the main reasons the standard format is needed. We need to have consistency in our process outputs, which means both a consistent execution of the process and a consistent format for the information that is the fuel for the process.

A tool is also a document that supports the process steps, but unlike the template, this doesn't result in an output to the process. In many cases, a tool may not need to be completed. Instead, it will provide guidance and recommendations on how the process step should be completed in the format of guidelines or recommended approaches—it facilitates the execution of the work. There will be some tools that can be completed—checklists or decision trees that we use to help us determine an appropriate management approach, for example—but they aren't going to become outputs. The output in this scenario will be the updated risk summary and risk management plan, which will have their own templates, but the tools, including the checklist and decision tree, have helped us generate appropriate content for those outputs. A tool should still be directly relevant to the specific step or steps in the process that it references. If it is a more general document on the entire organizational risk management approach, then it should be part of the support documentation—training material, reference guides, etc. We'll consider that section later.

As we create tools and templates, we need to be careful around wording. It's important to stress consistency in approach, but we have to be careful not to remove all flexibility from the people performing the work. If we become prescriptive in our approach, then we will not only cause resentment, we also remove the ability to interpret and apply the guidelines and by extension the application of skills and experience. Some of this is addressed in the processes themselves, but a lot of it is in the tools and templates that guide the application of process. Even relatively minor things can make a difference here. For example, I will never refer to a tool as *instructions*, instead referring to it as *guidelines*—a small, but not insignificant difference for the reader of that document.

When it comes to the creation of tools and templates, the first thing to determine is what we actually need to create. For templates, that's relatively easy. If it becomes an output, then it needs a template, either a new one for the step we are considering or an update to an existing one if the process updates or modifies something previously generated. For tools, there is a need for more subjective decision making, and this is another opportunity to engage the practitioner stakeholders to get their thoughts on the subject. We want to make sure that there are sufficient tools to properly support the process, but at the same time, we want to avoid making the people executing the process feel as though their role has been relegated to simply following checklists and filling out forms.

We also have to consider the starting point for the tools and templates. We considered a common look and feel in the process structure framework, so by now we should have a consistent document format—an overall design layout that clearly marks the document as belonging to organizational risk management. What we now need to decide is whether we are going to build our tools and templates from an existing document or start again from scratch. For example, even if the organization has never considered portfolio and program level processes in the past, it will likely have a project level risk management process in place, which will have templates for core documents—a risk management plan if nothing else. The likelihood is that the project level risk management plan will have many of the fields we need in our organizational process, so does it make sense to start with a brand new template or should we start from what we have? The answer will be unique to your organization, but the question needs to be asked rather than the answer assumed.

When we finally get to the details of the tools and templates, we need to make sure that all of the content is relevant and required. These documents need to support the overall concepts of effectiveness and efficiency, so we are looking for tools that are short and to the point and templates that are clear and contain only the relevant fields. Some guidelines for this include:

- Ensure that no field in the templates is captured unless it is part of the *static data* (e.g., portfolio, program and project identifiers, owner, risk ID, date) or is a field that is required to support the output—the real content of the risk. I have seen many templates that had long explanation fields that were never referred to by anyone consuming the template contents, and that's just creating work in the process that is not driving value.
- Create tools with the consumer in mind. These tools are going to be used as reference documents with people jumping into them when they need assistance. They need to be structured to allow for easy location of the relevant section—hyperlinks and bookmarks for example, and with easy to understand steps to follow—a decision tree that requires an hour of reading to understand how to execute it will soon be ignored.
- Try to avoid excessive use of free format fields. If your template is full of free text fields, it will be harder to retain consistency in the level and quality of completion. There will be some data elements that require this type of content, but where possible, the use of radio buttons and drop downs is preferred.
- Avoid assumptions of both knowledge and ignorance. I have seen a number of processes developed where the guidelines have become less and less complete as the process has proceeded. So in risk identification, the guidelines are extremely thorough, providing guidance on how to complete every field in a template, but by the time we reach adjust and refine, there is barely any guidance at all. We cannot assume that the same people will be executing each step, in fact, it is likely that they won't be, so we need to make sure that we provide the same level of guidance throughout the process. If we need to provide

guidance on where to find a project code in initiation, then we also have to provide the same guidance in adjust and refine. At the same time, we need to avoid assumptions of ignorance on the part of process practitioners. I have yet to find anyone who needed help from a set of guidelines on how to enter their own name in a template, and yet I repeatedly see it documented in the guidelines.

- Don't aim for perfection in this first attempt. At this point, we are trying to capture the main elements of our tools and templates while the inputs and outputs are still fresh in our mind. We will be revisiting these later on and can refine the details at that point.

This is an area where you need to engage stakeholders in the process early. I hear more complaints about tools and templates from the people executing process than I do on the processes themselves. Too often there is a disconnect between the process and the templates, and people feel as though they have to arbitrarily complete forms that do not advance the overall process in any way. Remember that we should always be aiming to ensure that every single action we implement in our process is supporting the achievement of the organization's goals and objectives through effective, efficient organizational risk management. The practitioners of the process aren't responsible for building the tools and templates unless they are part of the project team, but they can provide valuable input to the appropriateness of the material that is developed.

With organizational risk management, this is especially important. We are dealing with a strategic process, so it is likely that most of the people involved in the work are relatively experienced and skilled in their areas of expertise. Additionally, risk management is an area where specialist knowledge is often expected of the people doing the work—this isn't a frontline customer facing process. Therefore, while the point above about not assuming knowledge or ignorance is important, it is unlikely that any junior people will be engaged in many elements of the process, and the tools and templates should reflect that and be targeted at an appropriate level of user. Risk management tools and templates may well be more basic, as risk owners could be anyone in the organization, but the review and approval processes will require a degree of seniority and experience that should be reflected in their tools and templates.

What Are the Exceptions?

One of the aspects of process definition that is most misunderstood and most frequently excluded from the finished process is the concept of exceptions. Organizational processes are designed to provide a consistent approach that is optimized for the majority of situations, but the real world is not going to adhere to our process just because we've put so much work into it. There are going to be situations where the process simply doesn't work. While some of those situations cannot be foreseen, a large number of those exceptions are predictable and can be planned for within the process framework.

Think back to Section 1 of this book when we looked at risk management concepts. We talked about *known unknowns* for which we built a contingency reserve and *unknown unknowns* for which we had a management reserve. Think of exception handling within organizational risk management as the building of our process contingency plans—preparing for the known unknowns of process execution. At the same time, we have to recognize that there will be unknown unknowns that happen, situations that we did not foresee, and for which we did not plan. These still need to be managed through an exception handling process, but in this case, the specifics of the situation are more reactive—we didn't foresee the exception situation arising, and so we didn't have a contingency plan in place.

There are a number of categories where we have to consider process exceptions, and these are shown below. All of these are predictable scenarios and can have contingency plans developed for them. Even if the specifics of a situation are not foreseeable—a missed risk, for example, we can anticipate that there will be situations where risks will be missed, and we will need to deal with the situation. The broad categories for exceptions are as follows:

- Organizational areas that are structured differently than anticipated by our process. If an organization has a number of individual PMOs that do not share an integrated approach, then it is unlikely that a single common process is going to be able to be seamlessly integrated into all of them. Ultimately, there may be plans to combine the portfolio execution approaches into a single standard, or even merge the PMOs to create a single EPMO, but in the meantime, there will need to be some exception handling that needs to be considered. There is absolutely no excuse for missing these situations as they should have been flagged during process analysis and captured in our matrix (Table 20.1).
- *Fast track situations.* There will be scenarios where the standard processes simply don't work because of the urgency of the

situation. This may be a need to simply jump the queue for a particular risk or it may require certain steps or the entire process to be skipped because of the need to deal with a problem in a time critical manner.

- *Missed risks.* There will be situations where we fail to identify a risk at the initiation stage and may not even identify a potential problem until the risk has triggered. We need to be able to define how situations like this should be handled within the organizational risk management approach.
- *Process execution mistakes.* Processes are executed by people, and people make mistakes. There will be times when the flow of a risk through the management process becomes derailed due to human error, and we need to consider how we recognize and recover from these situations.
- *Constraint driven exceptions.* At various points in the organizational risk management process, there will be a reliance on limited skills sets—areas of the process where only one or two people have the ability to do the work. Inevitably, there will be points where either the work that needs to be performed is greater than their capacity to perform the work, or where they are unavailable for a period of time, resulting in either a need for an exception or a delay in the entire process.

As you can see, there are a number of different reasons why we may need exceptions to our process. Add to that the fact that these are exceptions, and it should be clear that we can't define an all-encompassing process for these. After all, if we could plan for them, then they wouldn't be exceptions! However, there are two things that we can do—define an exception handling process and define process additions within organizational risk management.

The exception handling process is not exclusive to organizational risk management. It should be standardized as much as possible across all of the organizational process groups and should focus on three things:

1. The approval of an exception—gaining approval to bypass the process in a specific situation
2. Documentation of the actions that were taken in lieu of the process being followed
3. A follow-up analysis of the process and situation to see if there are any process changes that need to occur as a result of the exception

In some situations, only the first point—the approval—will happen initially with the documentation and analysis being carried out after the fact.

This particularly applies when the request is to bypass the process because of a need for urgency—documentation and analysis will reduce the speed that was the whole purpose of the exception in the first place.

These exception scenarios may occur when we have high volumes of work, a shortage of resources, a set of urgent risk candidates, or some kind of similar situation. These are temporary scenarios and do not require a specific risk process to handle them; they need to allow for the execution of the process to be modified or ignored until the scenario is resolved. Instead, we should focus on ensuring that the application of the organization-wide exception handling process is documented for our risk management processes—who is responsible for sign-offs, regularity of reporting, and level of management monitoring.

When the exceptions are more fundamental to our organizational risk management process, we have to define one or more process additions as part of our overall process. Unlike the temporary situations described above, these are ongoing scenarios that may occur. We should look to build additional process steps when all of the following occur:

- The main process will not work effectively in the scenarios that are covered by the alternative processes
- The scenario cannot be changed or eliminated to allow for the main process to be utilized
- The scenario is likely to occur on a reasonably regular basis during the execution of the process

In some cases, these processes will be alternative paths through the process. For example, if a PMO uses a different project execution methodology from the rest of the organization, then there may be a need for a divergence within the process at different points to reflect a different sequence of events or set of inputs as a result of that methodology. The divergence should be as brief as possible with the processes converging again at the earliest opportunity. Wherever possible, the outputs should be consistently maintained across both paths to ensure that there are not downstream problems caused by inconsistent or unavailable data (an exception would be if an output is wholly contained within the divergent process steps). If your process has more than two or three of these divergences, then seriously consider whether the process is appropriate for the scenarios that require exceptions—you can't develop a standard organizational risk management approach for an organization that has multiple project execution models and expect to be effective. That situation requires the organization to look at the consolidation of project approaches.

In other situations, the exception processes will truly be additional steps that are only used in the case of an exception—the process for dealing with a missed risk, for example. These additional steps should be focused on understanding the impact of the exception and returning the item to the regular process as quickly as possible. In the case of a risk that has been missed, this might be an additional process step that involves a rapid impact assessment to understand whether the risk has triggered, the extent of the exposure, etc., followed by a return to the regular risk analysis process if the risk has not triggered.

This is another section where the inclusion of stakeholders will be beneficial. They can help with the identification of exception situations, which will again help them to feel as though they are invested in the process. Care needs to be taken here. There can be a tendency among people to protect their existing ways of doing things—pushing for exceptions to the new process rather than a compromise in their existing project execution approach—but the benefits of engaging stakeholders far outweighs the potential downside.

What Support Material and Process Is Required?

This final question to be answered in our journey from framework to detailed process involves the support infrastructure that organizational risk management requires. This needs to address obvious items like the process for training people on the process and providing them with support during the execution in the form of help documentation, as well as less obvious items like audit processes and the process for submitting improvement suggestions.

Wherever possible we should be trying to leverage existing organizational processes here. If the organization already has a process for training people on processes, then we simply need to provide our content into that process in the format that the process requires (using that process's templates) and use the well-established training delivery mechanisms that the organization has put in place. However, even if all of these elements are extremely well defined, we still need to make sure that organizational risk management is connected with those processes. The fact that we have a continuous improvement process in the organization is of no benefit if the people who are executing organizational risk management are unaware it exists.

There is still a substantive amount of work to be done in this section; even if the processes are well established, we still need to provide the process specific content. This will include the training material that needs to be developed, the elements that need to be audited along with the performance standards that are expected, and the user guides and FAQs. This is an area of the project where it will pay to bring in experts—a training manual is not just another document. It takes skill and experience to be able to write training material in a way that it can be readily consumed.

All of the material developed here should still be considered as a draft until the deployment process has begun (we will look at that in the next chapter). As the process is rolled out, there will inevitably be improvements and revisions that are identified before the final versions are ready for the entire organization.

Finalizing the Process

Once we have answered all of those questions, we will have the basis of a solid process, but we aren't quite done. We will have a lot of different process steps, each with their inputs, owners, and outputs. We'll have a collection of templates and tools, and we'll have a few (hopefully not too many) exceptions defined. Before we consider the process complete, we need to make sure that everything flows together logically and seamlessly. It needs to be a natural progression from the provision of project information that may contain portfolio level risks all the way through to the impact analysis and recovery from our triggered risks, and considering the adjust and refine elements that will inevitably be a key part of organizational level processes. Regardless of how the work has been broken down during the development phase, I always bring the entire team together to walk through the process in detail (you have to drill down to the lowest level for this) and it may take several sessions because it can be an intense and exhausting process. The team needs to review every step and ensure that all of the following are true:

- Progression from step to step is natural. We need to ensure that there are no overlaps between steps in the process, and that there are no gaps. Each step should naturally lead to the next step in the process without ever feeling forced or artificial.
- Every input is used in the right place. There has to be the potential to use every single input within the step for which it is an input. There may be scenarios where there is a list of decision support inputs and only some are used in a specific execution of the process, and that's fine, but if an input will never be used, it shouldn't be part of the process. Similarly, we have to ensure that there are no missing inputs in the process. If an item is not

explicitly defined, then it will not be an input to the process no matter how obvious it may appear during the process definition work. Finally, we have to ensure that an input exists at the point in the process that it is needed—it's not uncommon to find processes where an input is required earlier in the process than the step where it is created!

- Every output is identified correctly and in the right place. If an item is not created or transformed by the elements of the process step that it is tied to, then it is not an output of that process step and should be removed. Similarly, if an output is never used again in the process, then it may not be required at all and may be able to be removed (this isn't always the case, as we may be updating an organizational support document or producing an output that is used in another process within the methodology).

- Templates are complete and appropriate. Every field in the template should serve a purpose—either acting as an identifier for the risk, project, or program, or as a vehicle for carrying information through the process. If there are fields in the templates that are not used, they should be removed or the process should be updated to utilize the information. The templates should also be reviewed to ensure they are capturing all of the information that is delivered by the output of their process step. If the output includes risk exposure but the template has no field for risk exposure, you have a problem. The templates also need to be clear for users. If certain fields are only used when a template is updated or reviewed, for example, then make that clear. This is one element you should expect to modify during this final piece. Some of the other bullet points represent checks, but there is likely to be a fair amount of effort needed around templates.

- Tools are appropriate, complete, and assist with the execution of the process. There can be a tendency to produce tools just for the sake of it, or to make them longer than they need to be. You need to ensure that every element of each of the tools is adding meaningful value to the users of the process and that every tool is easy to use and interpret. If you have consolidated information from multiple steps in a single tool, ensure that the tool is clearly labeled for which information applies to which steps of the process. Again, this is an aspect where we should expect to make a number of refinements during the review.

- People responsible and accountable for each step of the process are appropriate. As you look at the flow of the process, you may find situations where the role responsible or accountable shifts back and forth several times in one area of the process. This may indicate there is an opportunity to consolidate responsibility or accountability to one person. This shouldn't be automatic, especially when we talk about responsibility (the person doing the work), because in risk management there will be a need for specialist skills at a number of key points and that may require frequent changes in responsibility. However, with accountability in particular, there are advantages to consolidating ownership if we can—it will reduce the risks of constantly changing ownership and provide continuity to the process. Remember there should only ever be one person accountable, so consolidating accountability to two roles is not appropriate.
- Exceptions are clearly and completely defined. This is never an exact science. There will inevitably be new exceptions that are only discovered once you start to roll out the process, but by running through the process end to end, you will be able to identify any areas where exceptions are likely that may not have yet been considered. In some cases, the organization's exception handling process can kick into gear, but there may be places where you need to define additional or alternate processes to ensure that as many eventualities as possible are built into the process.
- Support material is consistent with, and supportive of, the process. As noted above, this will be refined during the deployment of the process, but the closer that we can get it now to a final version the better. If the support material is incomplete or inconsistent with the process, then the process may lose credibility with the resources you are trying to roll it out to, and that makes it difficult to recover. You need to ensure that the training and support material aligns with the process—common terminology, clear guidance, even things like the same version of the process flow diagrams—the details matter. There should be an expectation that support material will develop and evolve during rollout and over the first few months of use as the level of organizational knowledge of an organizational approach to risk management evolves and matures. However, remember that support material also has to be able to support training of new employees who may not have any experience with a portfolio level approach to risk management.

After this work is completed, I like to return to something that we discussed in the last chapter when we looked at the process analysis. There I described building a matrix of the different departments, business units, PMOs, or offices impacted by our process development initiative and the process related items (process, tools, templates) that currently exist in each of them. We also added information on whether existing elements were reusable, had to be thrown out, or could be modified.

This provides us with a final chance to review the process and ensure that all of the gaps and inadequacies identified in the matrix have been addressed by the process that we have developed. If we identify matrix elements that still appear to be gaps in our new process, then they may represent missing process steps or missing exceptions. Or, they may simply indicate that a template or support document needs to be updated to reflect a scenario that is specific to one particular area within the organization.

The purpose of this check against the matrix is not necessarily to address all of the variances between the work that has been completed and the needs that were identified in the matrix, but rather it is to ensure that any gaps are conscious ones. We have made the decision to omit one scenario, group, or element because it does not easily fit within our process framework. In this scenario, we should document those remaining gaps, why they have been excluded, and a recommendation for how they can be addressed either in later phases of this initiative or in other projects that will align the group into more mainstream processes.

Once all of this work is done, then we can start to turn our attention to the next phase of the work and begin the process of implementation.

Web
Added
Value™

22

Process Implementation

At various points in this section, we have touched on the different models for implementing our developed process to the organization. There will be a lot of variables that go into the decision on how we actually rollout what we have built, and every implementation will be slightly different. One thing that I have found in the rolling out process is that it is always worth trying to keep a level of flexibility in the approach rather than locking yourself into an approach at the project outset.

There are some aspects that have to be determined up front. If your organizational risk management project is part of a larger portfolio level methodology deployment, then you need to operate within the framework of that program, for example. Some implementation models should always be rejected because the risks are too significant. I have been asked to roll out a risk identification and analysis process before the risk management elements were even defined, which is just asking for problems (try explaining to the people executing the process how the work that they are doing is going to help the organization to control risk in that situation).

In most cases, there are three distinct phases of developing a process implementation approach:

1. A preliminary plan during the project planning phase that will create some placeholder tasks in the project plan. This will be based on a high-level understanding of the project and the teams that will ultimately be using the process that is developed. Plan to avoid starting the rollout with a team that is inconsistent in its use of standard processes, or a group that seems to have a lot of exceptions. Instead, look for a more *mainstream*

group that can help avoid too many anomalies. At this point also try to determine how many phases the rollout needs to have—a big bang approach that implements every process element to every group at the same time should be avoided, but so too should the breaking down of the implementation into too many separate phases. That just results in the work dragging on and an extended period of inconsistency. At this point, the plan is more of a concept than anything else is, but it does help create a framework for the rest of the work.

2. A more detailed proposed plan that is created once the process development work is completed. That's where we are now, and this plan should be based on what we know of the process, our better knowledge of each of the groups, and our familiarity with the people in those groups. This will come from the work we have done with the stakeholders, as well as the people in our project team who may be from these groups. At this point, the plan shouldn't be expected to change much, but the only element that should be final is the approach we are going to take to prototyping or piloting the process. This will be based on a small group of users and will involve a closely monitored implementation—we'll look at that in quite a lot of depth later in this chapter.

3. A final plan based on phase 2 but that incorporates the lessons we have learned from the results of the pilot. This needs to be detailed enough to not only allow for accurate project management of the various steps, but so it can also be communicated to all stakeholders and they know when they will be impacted. This plan should focus on all elements of the process deployment—not just the process rollout itself but also training sessions, post implementation reviews, the start of process audits, etc.

I will never implement any process without first conducting a small scale pilot, so let's start our look at implementation there.

Determining the Pilot Approach

To get to this point we have invested a lot of time, energy, and money to build the best, most complete risk management process we can. The transition from development to implementation is the single biggest challenge that our project faces. If we get this wrong, then the process will be seen as a failure, and it is unlikely that we will ever recover. Our process is focused on risk management, and it's logical that we want to minimize the risk of such a failure occurring. The best risk mitigation process for any new process is to test the process with a small group and make adjustments before rolling the process out to all areas of the organization.

The pilot phase should be defined in terms of what needs to be achieved by the end of it:

- Confidence from the group of people in the pilot group that the process steps are complete and logical, that the tools and templates support the process as effectively as possible, and that the support material provides the right level of guidance and assistance with the execution of the process.
- Evidence that the process we have developed is more effective than the approach that was in use before—regardless of how formal that process was. This evidence should be as objective as possible and should focus on material areas—a reduction in total risk exposure (contingency reserve plus management reserve), fewer triggered risks, a shift from management reserve to contingency reserve (implying a greater number of identified risks and fewer unidentified ones), and more successful risk management (greater reduction in risk exposure due to management).
- A measure of process efficiency that is acceptable given the process effectiveness. The process may not end up being more efficient than the previous approach; in fact, that may be impossible if no formal organizational risk management was being undertaken (if you aren't managing risks, then you aren't spending time, effort, or money on risk management). However, any increase in risk management costs should be acceptable given the improvements in effectiveness delivered.
- An improved understanding of the most appropriate implementation strategy. This is the phase 3 that we discussed and should reflect both a sequencing of how the process will be deployed to different groups and an implementation approach that outlines how the implementation will occur.

Only when we are confident that all of the above have been achieved can we claim that the pilot has been successful and commit to a full scale deployment.

It should be clear that one of the most critical elements of the pilot is the selection of the group that will form this test group. With organizational risk management, we have both some restrictions and some opportunities in this area. The restrictions come from the fact that the process we have developed affects the organization at a strategic, organization-wide level—it's tough to find *the right* pilot portfolio if there is only one portfolio underway within the organization at any given time. That may force us to restrict the pilot to the program level, but even here,

---

Best Practice—Selecting the Pilot Group

The right group of people to pilot the process you have developed can make or break the project. Get it right, and you'll get meaningful feedback from real world scenarios that will allow you to improve the quality of the risk management process before it is rolled out department or organization wide. Get it wrong and not only might problems be missed, new issues may be introduced. Look for these qualities:

- A self-contained group—probably a program rather than a subset of the portfolio
- A group of people who, as a team, are experienced with project execution within the organization and with risk management, but who are open minded to change
- A group who collectively have the trust and respect of other members of the project execution resource pool
- A group consisting of individuals with a mix of senior and junior, new to the organization and long tenured, experienced with projects and new to projects
- A group including representation from a number of different areas within the department that the process will ultimately be deployed to (or from different departments if the process is being deployed organization wide during the first phase)
- A group prepared to challenge and question the process that they are piloting, but in a constructive and objective way
- A group with key stakeholders, especially sponsor and customer, who are supportive of the new process under development

You'll likely need to compromise some areas to get what you need in others, and the unique circumstances of your organization will impact how those compromises occur. However, the closer that your pilot matches to this ideal, the better your chances of a pilot that will drive meaningful improvements into the process. Remember, a successful pilot is not one that has no problems. It is one that identifies as many of the potential problems with the process as possible.

the options will be limited as the organization is unlikely to be starting a new program every week.

We may decide to wait until the next program is starting before we begin our pilot, and that has the advantage of allowing us to start a program *clean*—we start with risk identification on a brand new initiative and can see how our process works from end to end. However, this isn't a perfect approach; it will tell us nothing about how well the new process can operate alongside our existing risk management approach, so we don't know how a transition will work in an ongoing program. Even if we decide only to deploy the new process into new programs as they initiate (not recommended because it may mean we have to live with the existing processes for several more years until the last program that is currently underway finishes), we will be faced with two sets of organizational risk management processes at the portfolio level for an extended period. This approach may work if there is no process currently in place and we can introduce risk management quickly once the pilot has successfully concluded, but otherwise it is unlikely to be successful from an organizational standpoint. Conversely, this approach may be preferred by practitioners, as it avoids the need for them to work with two different processes concurrently.

These restrictions likely force us to consider a pilot within an existing program where we are faced with the need to deal with two risk management approaches in parallel. Theoretically, we could take all existing risks and enter them into our new process, but that is not only duplication of effort, it is unwise until we have a better level of confidence that our process is going to be successful. The entire concept of a pilot is to minimize the exposure that the organization faces until the process is tested and validated. This step may be taken at some point in the pilot, as a final verification, but not at the outset. Instead, we will likely use the new process for some or all of the new risk candidates that are identified, forcing people to work with two different processes, but delivering meaningful results for the organization more quickly.

I mentioned that organizational risk management also created opportunities, and that's caused by the fact that the process touches so many different areas of the organization. Whether it is determining impact, appointing a project level risk owner, or implementing a contingency plan with project impact, risk management reaches wider into the organization than nearly any other process. This allows us to engage a number of different organizational groups in our pilot and begin the process of socializing and communicating the process to different business groups. Of course, as we saw right at the start of the book, an opportunity is simply a positive risk, and we still need to manage the opportunity in order to maximize our chances of success.

Once we have decided on the approach we are going to take with the pilot, we need to implement the process with that pilot group, and that requires a hands-on approach from our project team.

Pilot Implementation

When we implement the pilot, our goals and objectives are different from when we are deploying the process across the organization. We should have an expectation in the pilot that the process is going to have problems and that changes are going to have to be made, both during the pilot process and certainly before the full scale deployment. The key is to recognize when those problems are occurring, to quickly determine the root cause, and then reinitiate the appropriate process development steps to correct the problem.

Many pilots are seen by both the project team and by the practitioners who are taking part as simply the first phase of implementation rather than one of the later stages of process development. This leads to unrealistic expectations followed by a negative impression of the process when it fails to live up to those expectations. Put bluntly, too many process implementations look at the pilot process as validation of the good work they have performed rather than discovery of the problems that remain.

The pilot, therefore, has to start with clear communication of the goals and objectives, along with ensuring that everyone involved has an accurate understanding of their role and responsibilities. Treating this as a separate phase of the project (or even a separate project in a program) allows for the definition of scope and the assignment of tasks just as any other project work package, and this all helps to create the context of this work still being part of the development initiative rather than the transition to production.

We need to try and ensure that the pure process execution elements of the pilot—the work done by the practitioners in applying the process steps we have developed—is as production like as possible. We use real projects and programs for the pilot so that situations and the pressures to effectively manage the risk are real. We use the same team members from those initiatives who would be using the existing risk management processes because they are not only the most appropriate people to speak to how the process works, but because they also have all of the other demands on their time that come from being part of a project, program, or portfolio. This is important in helping us to understand the impact of the processes on workloads and the time to execute process steps in real world environments where risk management is simply one of a large number of tasks that need to be performed.

At the same time, we need to have resources from our process development team closely monitoring how the application of the process is working. To get accurate results from the pilot, we don't just rely on feedback from the people applying the process steps, but also feedback from observers of that process. Observers are in the room while the group review of potential risks occurs, review of all of the completed templates to see how accurately and completely the work has been done, shadow people as they undertake analysis and prioritization, and perform countless interviews with people on how they find working with the new process.

This can be a difficult balancing act; we want people to execute the processes as close to the way they would in a normal production environment as possible, which will be difficult when they are being closely monitored. That means that we need to make sure the practitioners we use for our pilot are trained not just on the processes they will be applying but also on the role we need them to play in the pilot. The following are key themes to convey:

- Look for problems. The role of the people using the process in a pilot situation is to try and *break* the process, and they should be questioning every single aspect. They should be asking themselves at every step whether there is a better way of doing things, whether the template could be clearer or better designed, whether the support documents can be improved, and whether the flow is logical and consistent.
- Be realistic in the testing. We want people to try to find all of the problems, but they also need to focus on the scenarios that will occur in a production situation. By definition, processes are designed to deal with the 99% of situations that make up normal operations, not with the once in a lifetime scenarios that some people love to dream up.
- Separate execution from review. When they are executing steps in the process, the pilot group should be focused completely on those process elements and ignore any project people who are observing or monitoring the process. When the success or otherwise of the execution is being reviewed, then they are concerned with how the process can be improved. Another way

to think of this is to separate problems from solutions—when they are executing the process they are looking for problems, when they are reviewing what happened they are working to find solutions.

- Avoid comparisons with existing processes. Ultimately, there will need to be an analysis of the performance of our new approach relative to the process that is being replaced (or the lack of process), but that will be based on objective measures as much as possible. The people who are executing the pilot will be comparing an established, familiar approach with a new, imperfect process that is likely to be introducing significant change and with which they are unfamiliar; that's not a level playing field.

- Remain open minded. This is perhaps the most important instruction you can give to the people who will be piloting your process. As soon as the pilot starts, their colleagues are going to be asking them how things are going, what they think of the new approach, and whether it is *good* or *bad*. There is no problem with those questions being asked, or answered, but the team needs to understand that the process is still a work in progress—the things they are complaining about are the very things we are trying to identify and correct before the process is rolled out to a wider audience.

Clearly these activities are going to have some impact on the initiative the pilot resources are working on and for that reason we should ensure that the execution of the risk management pilot is included as a risk, or (more likely) a series of risks, within their program and projects. This risk is a nice way to start introducing the pilot group to the processes, as they can start to analyze and prioritize the risk and then determine a management approach and assign an owner.

We need to ensure there is a defined timeline for the pilot—this is part of a project after all, and schedule management is important. It can be tempting to continue the pilot for an extended period in order to build up the confidence of the people working on the pilot process, demonstrate an ability to fix problems, and prove the process works. That's not how we execute other projects, and it's not how we should execute this one. We should execute a planned set of tests as set out in the schedule and then analyze the results. We should have time built into the schedule for the corrective actions that will inevitably be required, and we should then be able to finalize the implementation plan for the complete rollout and begin that process.

In extreme circumstances, there may be a need to conduct a second pilot, but this should only be necessary if the project has experienced significant issues.

Organizational Risk Management Pilot Issues

The steps above can be applied to the piloting of any process development project, but we are focused specifically on organizational risk management, and there are some areas that commonly cause problems in developing a strategic risk management approach. You shouldn't consider the challenges identified in this section as the only problems that can arise—there is literally no limit to the issues that can occur, and like unknown unknowns, we can't plan for all of them. However, we can plan for the predictable problems. While the process development stage is where we would like to address these issues, there will always be some that slip through the cracks.

What follows are some of the more common problems you are likely to come across during an organizational risk management process pilot, the most likely causes of those problems, and the recommended solutions. Consider all of these in the context of your unique organization, but they should provide an accurate guide to the issues that you come across in your work.

Missed Risks or Missed Impacts

In any process, there is the possibility for human error, and it's possible people will make a mistake and miss a risk or incorrectly assess the impact of that risk. The likelihood of that happening is raised with a brand new process unfamiliar to the people executing it. However, if we are finding that there is a trend of things being missed, then something else is likely going on related to a problem with the process itself rather than simply an execution issue.

Cause

The most likely reason for this problem occurring is an incomplete list of inputs being identified. The fact that during the execution of the process we simply didn't recognize that certain risks existed, or that we didn't anticipate some of the impact areas, implies we were working with an incomplete picture—that some of the pieces were missing. It should be easy to spot a trend in the sources of risks that were missed; this will help us to narrow down the specific inputs that are likely to be missing from our process.

If analysis of the problem reveals that the inputs were identified but that they were not utilized because they were not readily available or the content was considered unreliable (especially where decision support inputs are concerned), then there may be a problem with the overall process sequencing. We may be trying to use information before it is ready to be used by the process step, or we may have the input too early in the process.

We also need to recognize that there may not be a problem with our process at all—it may simply be showing the shortcomings of other organizational processes on which organizational risk management relies. At the most fundamental level, it may indicate that the portfolio or program has failed to identify all of the initiatives that form part of it—a scenario that happens way more often than you might imagine.

Solution

If there are missing inputs, then the solution is obvious—add them in, making sure that they are at the right point in the process. Hopefully, we haven't missed anything as fundamental as an input that is part of an existing process, but there may be situations where the people executing the existing process know they also have to look at other items, even though they are not part of the formal process and we just never picked up on these informal elements. A more likely scenario is that, as we move risk management to a more strategic level, there is a need for additional inputs; and some of them were simply missed or added in too early—again a simple fix in the process.

Don't be surprised if there are a number of situations where the inputs that you need either don't exist or are not of a high enough quality to be useful. A number of the major, high-level inputs introduced to the process framework in Section 2—the organizational risk profile and organizational constraints hierarchy, for example—will likely never have been considered in your organization until this point. Your project is unlikely to have a broad enough mandate to be able to solve that yourself (you may be developing the tools but you can't force compliance), so you simply need to record it as a limitation to the effectiveness of the process until those pieces of information (and the processes required to generate them) are available. Also recognize that the pilot will likely be the first time that new organizational tools like the organizational risk profile will be used, and it will take time for people to learn how to complete it effectively and accurately.

A High Number of Triggered Risks

Risks are going to become real, and there is nothing we can do to avoid that. No matter how aggressively we manage the risks, it's not possible to eliminate every risk the organization faces. When we start piloting organizational risk management as a new concept, we should expect to see an increase in the number of triggered risks simply because we are giving more visibility into risk identification and management and are now aware of more risks than we were before this point—a case of better identification rather than worse management. That said, there will be situations where we find during the pilot that more risks are triggering, and we need to understand why that is occurring and address any process deficiencies that are facilitating that.

Cause

While there is any number of reasons why the volume of triggered risks has increased, if we determine that there is an underlying process issue rather than just a case of providing more visibility into previously hidden risks, then we should be looking at some areas as the most likely sources of the problem:

- Errors in the prioritization of the risks—either not having the significant risks prioritized high enough or not actively managing far enough down the priority list
- Errors in the determination of the management approach—making the wrong choice between mitigation, elimination, transference, and acceptance
- Errors in the application of risk management—likely resulting from a failure to recognize that the management approach that had been assigned was inappropriate or from a failure to recognize that the risk management activities likely need to be broader than was the case under the previous approach (because we are now focused organization wide rather than just project level)

We also need to recognize that this problem may be a further symptom of the problem that we looked at above around missed risk impacts. If we failed to identify the potential impact of a risk, then we will also have failed to manage for that impact.

Even once we have identified which of the reasons or combination of reasons above contributed to our problem, we still have to understand what led that to occur—the true root cause. For example, if the problem is that we didn't go deeply enough through the priority list to actively manage all of the risks that required active management, then the issue is likely one of resources—we haven't allowed sufficient people, time, or dollars to manage all of the risks that need managing. This is a common problem when we start to deploy a more substantial risk management process—the process improvements allow us to identify more of the risks to which the organization is exposed, but there hasn't been an increase in assigned resources (or not a large enough increase) to manage that increased risk list.

If the problem is more a case of inappropriately prioritizing risks or selecting the wrong management approach, then the underlying cause has to lay either in the analysis, in the interpretation of that analysis, or both. If the analysis is at fault, then this may be related to our problem above—issues with inputs in the analysis steps. If the interpretation is at fault, then that may indicate a training problem or a resourcing problem. The people performing the analysis should be experienced at risk prioritizing and determining the management approach, if only from project level risk management. If they are having problems correctly interpreting the analysis, then we should be looking at the additional information they are exposed to as part of our organizational process (they are unlikely to make mistakes interpreting familiar data).

You may well find that in many situations the problem arises from the team members having to prioritize and determine the management approach for more risks than that to which they are accustomed, resulting in them not having the time required for each individual risk to ensure appropriate prioritization and assignment of approach. We should have built a partial safeguard for this eventuality by allowing the assigned risk owner to question the management approach if they feel that it is inappropriate, and we have to assume that this safeguard has also failed, likely due to a reluctance to question a decision that had already been made.

Solution

As we saw in the cause section, there are a number of possible reasons for this problem, and not all of them require a process related solution to be implemented. In some cases, there may not be a real problem at all—the organization may decide that it can live with a larger number of triggered risks rather than invest more money in risk management. This is an acceptable strategy if the organization has a high degree of risk tolerance and/or if risk is low in the constraints hierarchy.

This may seem illogical—to invest in improved risk management processes and then not invest in the resources to manage the risks, but the greater visibility is still a tremendous benefit to the organization, as we are consciously accepting the risks rather than being unaware that the risks even exist. We also have to assume the lack of committed resources is a temporary arrangement and/or a reflection on the specific initiative, and this does not detract from the benefits the process can deliver in other circumstances.

If we assume that there is truly a problem that needs to be solved here—the symptoms aren't being caused by a conscious decision not to invest in risk management or by the same problem that we looked at above—then the solution is likely to be related to helping people to adjust to the new process. Specifically, it relates to how it differs from the project level risk management to which they are accustomed. This is actually good news—it means that the solution should be relatively straightforward and involve a combination of better training, better support documentation, and perhaps better tools. You should also look beyond direct process related training and consider the need for training in other corporate skills—handling increased volumes of work, the need for better prioritization, and better time management.

The most significant issue that you may need to address is related to the potential problem of someone executing on the process who doesn't feel authorized to challenge a decision made earlier in that process, as may be the case if the wrong management approach has been chosen and the risk owner hasn't questioned that decision. Organizational level processes are likely to involve people from completely different hierarchical levels, and every person involved has to feel empowered to challenge a decision made earlier in the process if they don't believe that the right decision has been made. Part of that can be addressed through training, but much of it will only be achieved through practical experience—a risk owner needs to see that he or she can question decisions that have been made without fear of criticism or retribution. The people whose decisions are being questioned need to recognize that it is not personal but rather is intended to ensure that the process is as effective as possible. This can be particularly challenging at the portfolio level where there may be a significant separation in organizational level between a project team member who owns the risk and the portfolio manager who signed off on a risk management approach.

Ineffective Risk Management

At the heart of risk management is the concept that the exposure the organization faces is reduced by the use of appropriate management techniques—that by applying specific actions aligned with some combination of mitigation, elimination, transference, and avoidance, we are reducing the impact of the risk, the likelihood of it occurring, or both. If that is not occurring during our pilot, or if the effectiveness of risk management is reduced compared with the existing processes, then we clearly have an issue that needs to be addressed. We saw in the previous scenario the possibility that the wrong management approach had been selected, but in this case, I want to focus on the right approach not resulting in the expected improvements.

Cause

Every risk is different, and every time failure occurs in managing that risk, there will be a number of different reasons behind that failure. As a result, we need to be careful in generalizing the reasons for management failure, and of course we shouldn't expect 100% success anyway. However, if the trend is indicating a reduction in risk management effectiveness, then it is likely the result of the risk owner not adapting their management approach to reflect the organizational nature of the risks. If a risk owner is used to managing risks that are fairly tactical in nature and isolated to just one project, then they will likely also have a fairly narrow focus in the way that they apply risk management techniques—monitoring focuses on the project, actions taken are focused on ensuring that the project impact is minimized and that the project is protected from that impact, etc.

With organizational level risks, this approach is unlikely to be effective because the needs of the portfolio, program, and project are unlikely to be perfectly aligned; and even if they are, the scale will be different, meaning that what may be acceptable at a project level is less so at the portfolio or program level. This will likely require compromise between the various levels, and if the risk owner is only focusing on the project level, then that compromise is always going to put the needs of the project above the needs of the program or portfolio—the exact opposite of what should be happening. The reasoning for this may be the result of a number of different factors:

- Insufficient training—the risk owner may not be sufficiently aware of the need to balance the needs of multiple different levels within the project execution organization

- Insufficient visibility—the risk owner may not have enough understanding or appreciation for how the risk can impact the program or portfolio levels
- Insufficient capability to act—the risk owner may not have sufficient ability to drive the changes needed to effectively manage the risk at the portfolio or program levels

In all likelihood, there is a combination of the above factors occurring, which will in turn require a combination of solutions to be applied. We also need to recognize that during the pilot stage, resources will be unfamiliar with applying the new organizational risk management skills they have learned. This is going to have a significant impact in the risk management area where the work is heavily reliant on skills and experience—on the application of process rather than simply the process itself. Until risk owners become comfortable, they will be more likely to make mistakes, and we should be careful not to overreact to something that may be nothing more than a lack of opportunity to practice applying their skills during a limited pilot.

Solution

Even if the issue is deemed to simply be a lack of experience with organizational risks, you still need to look for ways to resolve the issue prior to or during the rollout of the process to all users. While it's true that there is no substitute for experience, you can't sacrifice real initiatives for the sake of giving people experience in managing organizational risk effectively. Just like a problem with insufficient or inappropriate training, you need to try to find ways to provide people with as many opportunities to gain experience as possible before they are released to manage risks on their own. Some of that may be the use of practice exercises during training, and some may be the provision of project resources or risk management experts as mentors during the initial deployment period until risk owners are more comfortable with the management of organizational risks. This may require a lengthy commitment—any project level resource could end up being a risk owner, and they will all have some degree of learning curve until they are comfortable with the process and their ability to execute. This may require us to build an *ask the experts* type function into our process support material—an e-mail address monitored by risk experts in the PMO, for example.

The nature of the training we deliver also needs to be considered here—this is not simply a new process that people need to be trained to use; this is a fundamental shift in the way risk management is executed within the organization, and it requires an adjustment to the approach that risk owners take to executing the role. Training can handle some of this—providing people with the skills that they need to take a more strategic view, to balance conflicting priorities, and to consider additional variables in their decisions and actions, but that training needs to go beyond pure process training into more soft skills and leadership type skills development.

In other cases, there may be a need to review whether the process for determining the appropriate risk owner needs to change—the owner that can effectively manage a project level risk may not be the same person as the owner that can manage a program or portfolio level risk. However, we also have to consider that if we appoint a more senior resource to manage a risk, then they are likely to be a little too far removed from the project level (where impacts and warning signs generally appear first) to be able to effectively monitor what is happening. This may require us to look at supplementing our process approach with one or more of the following:

- An additional *supervisory owner* type of role. Appointing multiple owners to risks creates the tendency for both parties to assume that the other is handling it, but having an owner at the project level with someone else monitoring to ensure that the program and/or portfolio needs are met may work. Similarly, if most of the impact is at a higher level, an owner at the program or portfolio element of the work may make the most sense with someone monitoring and reporting on symptoms and early warnings that may be occurring at the project level.

- A formal review process that we build into our risk management activities. This will continue the concept of a single risk owner, simplifying the management process compared to the approach described in the above bullet point, but will add a documented regular review of the management actions taken. This may be nothing more than a weekly review of risk summaries to ensure that risks are being appropriately managed, or it may be a more active review involving discussions and active checks of work. This formal review works well for critical risks, but is likely to be an expensive approach for all risks, and may lead to risk owners feeling as though they are not trusted to act—even if the focus is on ensuring that the right activities are taking place, people will feel it reflects on them personally.

- A built-in ownership review process after the risk owner has been managing the risk for a period of time. Whereas the previous bullet point focuses on a review of the work, this is concerned more with a review of the suitability of the owner that has been assigned. Not looking at the performance of the owner, but rather on whether the owner who was originally appointed is actually the most appropriate for the risk in question given what we have learned about the management of the risk during the initial days and weeks that the risk was being managed.

Beyond any process driven solution to this problem, there has to be recognition that the actions associated with risk management require more than the execution of a series of established steps. The process is vitally important in establishing the right steps that need to be applied consistently in order to maximize the likelihood of success, but the individual risk owner needs to be able to apply judgment to the unique circumstances that each risk presents. The best process in the world cannot anticipate every scenario, and the most effective risk management will come from the right owner making the right decisions at the right time—the process can simply maximize the likelihood of the right owner being identified and empowered with the tools, skills, and experience needed.

Failed Contingency

Risk management is not an exact science, and risks will trigger no matter how well the risk management processes have been applied. That's why the risk management process includes the development of contingency and the ability to apply that contingency immediately when it is needed. When you pilot your risk management process, you need to make sure that there are sufficient risks included in the scope to test the effectiveness of contingency plans. If a risk process is rolled out before the contingency related elements are perfected, there can be significant impact on the ability of the portfolio to achieve its objectives—far better to deal with a greater number of triggered risks with solid contingency than a smaller number of risks with ineffective contingency.

If your pilot indicates that the contingency plans developed as part of the process were ineffective, then you need to identify the cause and address the problem before you begin the process rollout.

Cause

At the simplest level, there are only three possible causes for contingency to fail:

- The contingency plan developed was inadequate or incomplete. Either the process and/or template wasn't rigorous enough to define a comprehensive contingency plan, the plan wasn't enhanced and revised as risk management was executed and more was learned about the unique risk situation, or the risk owner was incapable of developing an effective contingency plan.

- The contingency plan was not implemented effectively. This may have been as the result of a faulty plan that led to the wrong steps being undertaken, or it may have been caused by resistance among the impacted areas to the contingency steps that the risk owner was trying to implement. It may also be the result of a panicked response to the triggered risk that resulted in a lack of focus on how to limit the damage.
- The risk trigger event(s) were not recognized or acknowledged in a timely manner. This is frequently the biggest problem and is the result of a mindset among risk owners that a triggered risk represents a failure on their part. As a result, they either refuse to accept that the risk has actually become real, or they think that they can somehow *un-trigger* the risk through continued management.

In each of these situations, you may have underlying problems with the processes themselves, or with the tools and templates related to those tools. That's likely to be the case if there seems to be a lack of clarity around what needs to be done—confusion is a result of the absence of clearly defined steps to take that are at a sufficient enough level of detail that they can be applied almost automatically in a risk triggering situation. If the issue is more associated with how the contingency should be applied, when to implement contingency, or why portfolio resources have to adjust their work because of the demands of a risk owner, then the issue is more likely related to the issue of insufficient training and experience. It is also possible that the process doesn't allow for rapid enough escalation of the issue to portfolio and program leadership, or that those leaders don't become actively involved quickly enough.

Even where risk owners are experienced at implementing contingency, there can be a reluctance to act decisively with portfolio level risks because the impact may be felt far beyond the project level that they are more used to handling, and there is the fear that they may make a mistake and cause unnecessary disruption across the portfolio.

Solution

In some ways, the practical steps to solve this problem are no different from the solutions we have already considered. We need to ensure that the processes and the associated tools and templates are rigorous enough to guide a risk owner to develop a comprehensive contingency plan that can be applied when needed. There also needs to be a review task built into the ongoing risk monitoring to ensure that the contingency plan is maintained as the risk evolves and changes.

We also need to make sure that the training and support material are robust enough to provide risk owners with the confidence they need to decisively execute a contingency plan as soon as a risk triggers. This training needs to extend to everyone involved in portfolio execution because they may find themselves impacted by a contingency plan even if they have nothing to do with risk ownership. They need to be prepared to deal with the potential disruption that can occur when contingency is implemented and be prepared to implement their own recovery tasks if necessary.

However, the solution to contingency problems goes beyond improving process and training—it goes to the cultural environment that exists within the organization. You can't always identify cultural issues until the process is rolled out across the organization, especially if they are isolated to certain areas within the organization, but contingency is one area that you should be monitoring closely for these issues to occur:

- Consistent reluctance by a risk owner to implement contingency when a risk triggers. This implies that the risk owner does not feel secure enough within the organization to make the call to implement contingency without repercussions. It may be the owner is afraid to be branded a failure, or that he or she will have difficulties working with the groups impacted by the contingency work, but it is an indication that the process has not yet been fully embraced within that part of the organization.

- Frequent problems in effectively implementing contingency in one or more areas of the organization. This suggests that one or more groups is either not prepared to shift from their own priorities to implement contingency steps that are designed to maximize the ability of the portfolio to meet its goals, or that there are frequent questions or challenges around the actions involved in contingency. Again this indicates that there is a lack of trust and acceptance of the overall organizational risk management process that is leading to second guessing of the work around each contingency plan—putting individual needs ahead of the greater good of the portfolio's needs.

While these situations can be solved on a case by case basis, this will only ever be a temporary solution that will need to be repeated every time that a risk triggers. The better solution is to identify and address the underlying issue that is resulting in a refusal to accept and embrace the organizational risk management process. This may be a lengthy process and may lead to a change in the deployment schedule, but it is necessary if the process deployment is going to be ultimately successful. Review Chapter 18 for some ideas on how to assess the environment and culture and focus on demonstrating how the process that you have developed helps the organization as a whole. Also, be prepared to accept that they may have a valid concern that needs to be addressed, if only as an exception process for their unique situation.

Process Rollout

Once you have completed the pilot and have addressed all of the issues identified during that stage (testing and validating the changes where necessary), you are ready for the rollout across the entire organization (or the area of the organization that is within scope). This is step 3 in our three-step process outlined at the start of this chapter.

At this point, you should have a final plan that is not subject to any further planned changes. This is a project in itself and should have a detailed plan that documents each of the steps involved. Your plan should have been built considering all of the variables:

- Speed of deployment. This is an organizational process and so has extremely broad scope. Even if it is only being deployed to one or two PMOs in a much larger organization, there are going to be a number of initiatives and people impacted, and until the process is fully rolled out, there will be nearly guaranteed confusion. Some projects will be using only their historic risk management approaches while others will be considering portfolio-wide risks, which will inevitably have impact on projects that aren't using the approach. This will reduce the efficiency and effectiveness of both processes and may lead to resistance in accepting the new process; it will also increase the risk exposure for every initiative that is underway during the rollout period. For these reasons, I always try to complete the rollout in a relatively short period so that the period of overlap will be minimized. However, that must consider the other points.

- Capacity to absorb the process. Rolling out the process is going to be time and effort intensive for the areas of the business that are impacted by the change. There is a need for communication around the upcoming changes and the provision of opportunities to voice concerns and ask questions. There will also need to be training on the process, and that will need to be varied for different people involved in different elements of the process. Once the process is in place, there will need to be some hands-on guidance from the project team, which will slow down the speed of execution, as will the fact that people are dealing with unfamiliar processes. Finally, there will need to be formal and informal opportunities to provide feedback and commentary on the rollout. All of this will take time and effort away from the work of project execution, so the rollout will need to be throttled to ensure that the impact on ongoing initiatives can be absorbed. There's no humor in the irony of a portfolio that fails to meet its objectives because too much effort was diverted toward implementing an organizational risk management approach.
- Capacity to deliver the process. In just the same way that we have to consider how capable the organization is to consume the new process, we have to consider how ready the organization is to deliver. There should be preliminary work done on initiatives like train the trainer (the process of training the people who will then be responsible for training all users) to increase the number of people who can assist in the rollout and prepare roles like corporate training and help desk, but there also needs to be consideration of how thinly the project team can be spread to rollout the process. There needs to be sufficient levels of resourcing to ensure that coaching and support can be given until people become comfortable with the new approach, and there will need to be people available to receive and respond to questions and concerns. Even logistical issues like the availability and capacity of training rooms needs to be considered in order to ensure there are no problems with the rollout.
- Portfolio execution cycle. If your organization operates on an annual planning cycle, then there will be times in the ebb and flow of portfolio execution that are more naturally suited to the rollout of the process. It may make sense to try and complete the rollout before the launch of all of the newly approved

initiatives, or you may want to defer the rollout until after the current surge of new risks requiring analysis has died down. If the organization doesn't follow an annual cycle and has new initiatives launch throughout the year, then you may be able to plan rollout on a project team by team basis based on the completion of the project lifecycle.

- Variations in the groups to whom the process will be rolled out. You will likely have some groups that are more familiar with formal risk management than others, and you may even have groups who have used some previous organizational level processes. These groups are going to need a different type of deployment than those that are completely new to the concepts. In one case, you have an understanding of the fundamentals of risk management but also the need to *unlearn* an existing approach; and in the other, you need to start with risk management basics but have the advantage of being able to deploy into a *pure* environment.

Of course, plans never go smoothly, and you will need to adapt the rollout as you face challenges and delays, but at this point, you should resist making changes to the process itself. The pilot should have resulted in a process in which you have a high degree of confidence. While there will inevitably be further evolutions to it after deployment as continuous improvement begins, if you start changing the process on the fly during the rollout, you will lose control of what is happening, end up with variations in the way that the process is implemented, and create confusion both in your project team and in the groups utilizing the process.

Once the process has been deployed to all of the groups who will be using it, you can start thinking about shifting from project mode to ongoing operations mode. This will involve the handover of the process to the PMO group that will ultimately be responsible for the management of the process and the ongoing evolution through continuous improvement. This shouldn't be an immediate switch. While some of the project resources will be from the PMO and may well form part of the ongoing ownership function, there still needs to be a transition period to ensure that the PMO is capable of fulfilling all of their support functions—training of new resources based on the material that the project team has developed, integrating the audit tasks into the existing project audit processes, and adding all of the templates and completed artifacts into the project library.

Project Closeout

The final stage of the project is of course the formal closeout. In most aspects, this is no different from any other project, in that the lessons learned exercise will identify ways that project execution can be improved in the future. However, if our project is part of a larger initiative—the expansion of the process to other areas of the organization and/or the development of additional portfolio-wide processes as part of a larger methodology—then there are some specific items that need to be captured and communicated to the other projects within that process development program. The specifics will vary from program to program but may include some of the following:

- Repeatable subprocesses. There may be specific process elements that can be reused, almost in a cut and paste capacity elsewhere in the program. For example, the process you have developed for formal reviews and approvals may be able to be lifted directly and placed wherever similar reviews are needed in other processes. Not only will this reduce the amount of process development work that is required, it will provide consistency across the different processes, which will make the processes easier to adopt and will give the sense they are part of an overall cohesive methodology.

- Templates for process artifacts. Although the tools and templates used for each process element are unique, there are common elements that can be leveraged to provide the same advantages as those described for the subprocesses. Common template layouts for things like project codes and identifiers, sign-offs, and checklists will again convey a sense of a common identity and make adoption easier.

- Glossary of terms. If your project is the first element of a major program, then you have developed your own set of terms and definitions, even if that wasn't done consciously. Some of those terms will be common to your organization, but others will be specific to the processes that you have developed. By completing a glossary of these terms that the next wave of projects can use, you are again contributing to consistency.

- Resource recommendations. You have worked with a number of different people in the development and deployment of the processes, and you will have formed opinions on the suitability of each of those resources for inclusion in future initiatives.

Clearly, this is a sensitive area, but if you have had to deal with people who are unable or unwilling to contribute to the process, then this is important information to provide (confidentially) to the next projects within the program. You don't want a team of people who are reluctant to challenge and question the approach, but you do need to ensure that those challenges are constructive and contribute to the development of better processes for the organization.

- Exceptions. We talked in Chapter 21 about the need to develop exception processes, and these will have identified areas of the organization where the standard process may not naturally fit for any number of different reasons. By ensuring that these areas are clearly communicated, we can either help future process development projects to prepare for those exceptions, or potentially implement a new project (or change to an existing project) to try to standardize those organizational areas and eliminate the need for numerous exceptions.

Once all of these are documented and are captured as part of the larger program documentation, then the initiative can be shut down and resources released back to their operational areas or to their next assigned project.

23

Process Improvement

The rollout of the process within the organization represents a major milestone, and while there may be many more initiatives that need to be executed before an organization-wide portfolio execution methodology is in place, the achievement should be celebrated. However, the work is not done. We need to ensure that the implemented process is being accepted and executed properly. We also need to ensure that there are processes in place that will allow the process to evolve over time as the organizational needs and priorities shift.

That's where process review and improvement comes into play. Even if your organization already has a process review approach in place, there are some unique challenges associated with organizational processes that need to be addressed. I want to look at two distinct review processes in this chapter, starting with a process that validates the implementation and corrects any immediate issues.

Organizational Implementation Review

After the process has been deployed, we need to conduct an initial formal process review driven by the PMO that will be assuming ownership of the process. This should be carried out as soon as there are sufficient data points to provide an accurate review—certainly within the first portfolio cycle, and for things like risk identification and analysis it can start within a few weeks of the process going live. It is important that this review be driven by the process owner rather than be in the form of a post implementation review conducted by the project team because it helps to establish the ownership for the practitioners. They immediately associate accountability with the PMO, and that will be important when it comes to process audits and performance reviews. It also provides PMO resources with a great learning opportunity to become familiar with the process.

However, the purpose of this initial review is very much in the form of a more traditional post implementation review that might be conducted by a project team, and project resources will need to provide the PMO with support during its execution. The review is designed to identify and address the following:

- Problems with the initial acceptance of the process. This may be in the form of practitioners continuing to use the old processes or continuing to use an unstructured ad hoc approach. It may be that people are trying to follow the processes but are facing pressure from higher level stakeholders—project managers pressuring their resources to focus on their project deliverables rather than spend time on portfolio level risk management tasks, for example. We need to understand the underlying issues that are driving this behavior and address them before anyone can think that the new approach can be ignored.
- Problems with the understanding of the process. This may manifest itself through mistakes in an element of the process— e.g., not following the process steps correctly, incompletely or inaccurately completing templates, or not using all inputs. The symptoms may simply be points of delay in the process—tasks that are taking longer than planned, bottlenecks, or apparent shortages in the number of resources available for the work load. In many cases, there may not even be awareness among practitioners that they are executing the process in a way that is different from what is intended—they may believe they have accurately understood the instructions, guides, processes, or templates.
- Confusion within the organization. No matter how much work has gone into preparing the organization for the process rollout, there may be individuals or groups of people who are unaware of the implementation or have misunderstood the impact on them; this can lead to confusion over how they should respond when they have to deal with the process for the first time. This can easily turn into resentment or a perception of having been excluded or forgotten during the deployment.
- Implementation gaps. Once the process has been rolled out to the organization, you may find there are pockets where the

process has not been deployed and where that gap is preventing the process from being optimally effective. This may have been an inadvertent omission or a conscious decision not to deploy to the impacted area yet, but steps need to be taken to address any lack of effectiveness, even if that means making changes to deployment schedules, creating temporary *bridge* processes until a full deployment is complete, or a more unique solution for a specific scenario.

If the implementation of the process is occurring over a period of time or if it's being introduced at the start of new initiatives, for example, then this review should also be phased. By conducting this implementation review for the early implementations while later implementation phases are underway, the PMO can take steps to address systemic problems that are identified with the implementation approach before they are proliferated elsewhere in the organization. This obviously requires close cooperation between the PMO and the project team so that the implementation plan can be adapted effectively and efficiently.

Before we look at that in more detail however, we should note that not all situations identified during the implementation review will need to be addressed immediately—in fact, some of them shouldn't be addressed straightaway. There will be situations where the best approach is to simply put a potential problem on watch, to review it more broadly across the organization and in more depth before doing anything. This will allow the PMO to see whether changes have to be made to the process or support materials, whether the problem is simply the result of people going through a period of adjustment to the new process, or there is a unique situation in one part of the organization. If further analysis of these scenarios does reveal that a change to the process or support materials is required, then this can be rolled into the same review and implementation process as the ideas that are generated during the continuous improvement processes. Those are covered later in this chapter.

Where the implementation review highlights issues that cannot wait to be addressed, then there needs to be careful consideration given to how to address the problem from both a practical and a perceptual perspective:

- The process changes need to be made as quickly as possible in order to prevent further problems as more risks go through the process. At the same time, there needs to be minimal negative impact on the projects, programs, and portfolio that are using the process. This may be difficult if the flawed process has driven potential problems into those initiatives. This is another situation where the organization needs to make a risk based assessment. The flawed process has introduced a new risk into the initiative, and the impact of that risk needs to be assessed to see whether any re-assessment or similar changes are needed to the risks that have been through the process.

- There needs to be a balance between making the changes to the process and maintaining the rollout schedule to other organizational areas. In some cases, there may need to be a delay in the schedule in order to incorporate the changes; however, this shouldn't be the automatic reaction. If your organization has no formal portfolio level process, then organizational risk management may still be driving significant benefits, even if it isn't perfect. On the other hand, it is less disruptive, and more effective, to implement the right process the first time rather than have to make changes as soon as the process is implemented.

- The reaction of process practitioners cannot be ignored. Even if the process is broadly welcomed by the people who will be executing it, the implementation will still be disruptive; for most people involved, it will be driving additional work into their already heavy workloads. The organization doesn't have multiple attempts to get the process right without losing credibility, which requires a balance between allowing a flawed process to continue and making sure that the change implemented is the right one. Continuous improvement as a concept tends to be welcomed as a concept because it is a slow evolution. The changes that happen as a result of an implementation review tend to be more substantial and more rapid. If they create multiple cycles of those changes, the process can quickly lose credibility among the people who need to execute on it.

Taken together, these points mean we are faced with the need to take decisive action quickly, but it needs to be the right action; otherwise, we damage the process. That's not an easy thing to achieve, and it requires considerable focus. Hard and fast rules are difficult to provide when there are so many unique variables in every organization, but there is one that is close. If it appears as though the implemented process has a fundamental flaw that cannot wait to be addressed, then the PMO should put all of its efforts into identifying the source of the problem, correcting it, and deploying the fix. If the project team is still in place and deploying a later phase, then they should be brought in to assist—there's no point in keeping to the rollout schedule if it means a flawed process is being deployed. If the project team has been disbanded, we should try and secure time from the various subject matter experts who were part of the project team to identify and solve the problem.

All of that said, this will only be a serious issue if there has been a fundamental flaw in the process development project and pilot that led to this being missed. In the vast majority of cases, what may at first appear to be a significant issue will end up being a misunderstanding, a misinterpretation, or an exception to the process. All of those are serious and need to be addressed, but they aren't issues that require a shift in the entire process.

Continuous Improvement

The concept of continuous improvement as an effective way to improve processes is well established and well documented, and there is no need to reinvent that wheel here. However, the continuous improvement approach that you likely have in place for project level processes may not be sufficient to support processes that are wider reaching. When we move to an organization-wide process, there are a number of additional complexities:

- There are vastly more scenarios in which the change needs to be validated. If we are proposing a change to a project level process, then we can test it out in a couple of places and have a reasonable level of confidence it is going to deliver the improvements we expect. With organizational processes, we face many more scenarios that are impacted by each process element—portfolio, program, and project levels; multiple different areas of the organization; and countless unique scenarios. As a result, we either need to undertake a significant amount of testing of the change or make a dangerous assumption that a small scale test will be valid across all different levels and organizational areas.
- Communication and rollout of changes are more time consuming. With so many different areas of the organization utilizing the same processes, the communication of upcoming changes; the answering of questions and clarification of areas of uncertainty; and the physical deployment of updated process documents, tools, templates, and support material will take a lot of effort and weeks or months of elapsed time. That will result in increased potential for confusion with multiple versions of the

process in use at the same time and risks on the same initiative being managed differently because of the delays in rollout.

- Potential for areas to be missed. With the less formal and less structured rollout that inevitably occurs with continuous improvement changes, there is a possibility that some organizational areas or individuals will be missed or that they will choose not to adopt the change without the PMO that owns the process recognizing the disparity. This not only creates inconsistency, it has an impact throughout the organization in everything from acceptance to audit.

This doesn't mean we can't utilize continuous improvement for organizational processes—the concept is just as valid as with any other process—but we do have to modify the way that we capture ideas and implement changes. I have found that because organizational processes are fairly new (relative to project level processes) and because they are likely to engage a large number of people from many different areas of the organization, there are actually a lot of ideas generated for how the process can be improved. However, I have also found that many of these suggestions are from a narrow perspective—they are ideas that make sense to the person suggesting them and their limited exposure to the process, but they don't stand up when tested across the entire scope of the process.

This is a tough balancing act. We want to encourage suggestions for improvement, especially early on in the process's lifecycle when the opportunities to improve the process are the greatest, but we don't want to waste time investigating suggestions that don't make sense in anything other than specific situations. I find the most effective and efficient way to tackle this is to have a regular cycle for reviewing and implementing continuous improvement driven changes. Technically, I guess that means the improvement isn't continuous because we hold them and then review as a group, but it doesn't cause practical problems as long as the expectations are set that suggestions may not be reviewed and responded to immediately.

Review and Implementation Process

Ideas for process improvements that come through the process improvement process can be captured using existing processes—the same forms, the same submission process, etc. There is no need to overcomplicate things. However, the submissions aren't subjected to a detailed review when they are first submitted. Rather, once they are given a preliminary *sanity check*, they are collated until we are ready to initiate the next instance of the review and implementation process. This process is shown in Figure 23.1. This process also applies to any of the non-immediate change recommendations that come out of the organizational review, although in that case there may not be a need for the various updates to the submitters of the proposed changes because they will have originated from within the PMO's own review process.

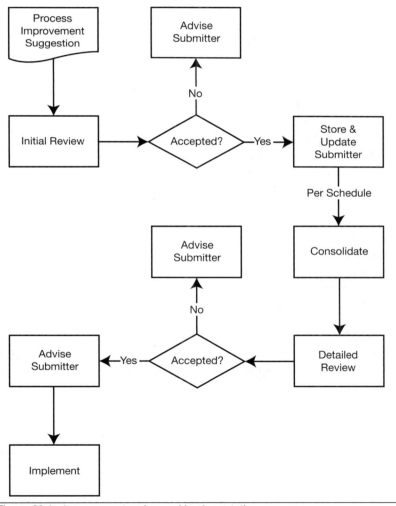

Figure 23.1   Improvement review and implementation process

The schedule for this process will be driven by the organization's overall approach to portfolio execution. If the organization operates on an annual planning cycle, then an annual review cycle that implements the process changes shortly before the initiatives associated with the new annual portfolio launch makes sense. If portfolio elements launch on a more even cycle throughout the year, then the schedule can be developed to suit the capacity and convenience of the PMO. You should deploy the changes consistently across the entire organization rather than to phase the deployment. It will help to avoid the confusion that can be caused by having multiple different versions of the process in use at the same time.

Hopefully, your organization's continuous improvement process includes steps to keep the original submitter of a suggestion informed on the process of their idea, but it's an even more important step here due to the inherent delays between the submission of the idea and its comprehensive review by the PMO. It's important to not only acknowledge that the idea has been received but also to set expectations that it will be reviewed as part of a scheduled review process and may end up being consolidated with other ideas.

Consolidation is necessary not only to make the volume of submitted ideas more manageable, but also to align similar ideas. With organizational processes, there will be great variation between the experiences that each practitioner has, even if they are performing the same function, simply because the reach of the process is so considerable. That will result in different people identifying different perceived challenges with the same process element, and a comprehensive review of the process needs to consider all of these different perspectives. It's important to note at this point that this is a case of *perception is reality*. If there are consistent concerns expressed with an area of the process, or if a number of people feel that similar improvements are required, then you likely do need to make some change. That change may be related to training or support documentation rather than the process itself, but you should be focusing on trends in the feedback, especially with newly implemented processes. This is another advantage of the scheduled periodic reviews—it helps to support the identification of trends by grouping suggestions together.

When you are conducting the detailed review of the process improvement suggestions, it's important to ensure that the review considers all situations with which the process has to deal. When we developed the process, we considered the need for exception processes to deal with steps that had to be adjusted for specific scenarios, and you should be on the lookout for these in the early reviews of improvement suggestions. If you find that one particular process element has the vast majority of suggestions coming from just one area of the organization, then this is almost certain to be a scenario where an exception process is needed and was missed.

Any changes that you do decide to make as a result of the review should be treated just the same as when the process was first developed—any change in the process may drive changes in inputs or outputs, and in the tools, templates, support material, and training. Even if the change appears to be the simple addition of one more information source as an input, there may be an effect. Does an upstream process element have to be adjusted to ensure that quality information is available as an input or are there other process areas where the input should also be added?

Regardless of whether any changes are made, there should again be clear communication with the original submitter of the suggested improvement to ensure that they understand the decision that has been taken, as well as the reasoning behind the decision. If the decision is to implement the change, consider making that the person or people responsible for the suggestions that drove the change are briefed on exactly what is changing and be given the opportunity to comment before the implementation begins. That can not only help to ensure that they have a full understanding of how their idea has developed and evolved, it can also turn them into champions of the change with their colleagues. However, this should not be seen as an opportunity for them to make further changes—the change is the output of a thorough process development process, and you need to be confident in its capability to improve the process.

When the changes are rolled out to the organization, it is vitally important that there is accurate versioning of the process documents—a date of last change is not on its own sufficient. There has to be a specific identification of it being the current version to avoid confusion over when the newest version was produced. It's inevitable that different areas of the organization will maintain local copies of documentation, even if the process defines the location of the *master* version. You need to ensure that there is an easy way to compare the documents in use with the version that is in the official repository—preferably for practitioners, but, if necessary, for auditors.

24

The Impact of
Technology

We have spent a substantial portion of this book developing and implementing an organizational risk management process and barely made a mention of how technology integrates and supports the risk management process. This may seem odd; after all, computers are an integral part of everything we do in our lives, and it's likely that technology will have a major part to play in your process development work. However, it was a conscious choice not to talk about technology until the end of the book, for a number of reasons.

First, we don't want to imply that a certain technology standard is necessary in order to be able to develop and implement effective organizational processes. Some of you will have access to sophisticated portfolio management software with powerful workflow engines that can automate many of the process handoffs. Some of you will simply use a shared drive to store project documents and rely on simple diagramming functionality within your office productivity tools to draw process flows. Both models can be successful—the specifics of the process will simply be different in the two situations to reflect the different technology environment.

Second, and most important, success in organizational risk management (or any other process) requires us to be focused on the right things, and that means the processes themselves. We should view technology as an enabler and facilitator that can make the execution and management of processes easier, but it should never drive the process discussions. Put simply, we should never compromise the right approach to make it easier to fit into a tool we want to use.

Now though the hard work of developing process is done, and we can consider some of the ways technology can assist with the management and execution of the work we have completed. We're not going to look at how collaboration tools can be used to facilitate multiple people working on the same document—that's fundamental to such tools, and any number of user guides on the specific tool implemented in your organization can help you with that. Instead, I'm going to focus on the piece of specialist software that is likely to have the biggest impact on organizational risk management—project portfolio management, or PPM, software. If your organization hasn't invested in such a tool yet, it is worth considering—not specifically for managing a risk management process, but because if your organization has a mature enough project execution approach to be able to implement organizational risk management, then it is also likely to be able to gain significant advantages across the portfolio from such a tool.

Risk Management and PPM Software

In recent years, PPM software has become much more accessible. The use of modular elements has made implementation easier and has allowed the tools to grow with organizations. The evolution and acceptance of cloud infrastructures has made the tools more accessible to organizations that don't have the ability to host and maintain the applications in house. Software as a service (SaaS) models have made the cost much more reasonable and have eliminated the need for multiyear commitments.

At the same time, the tools themselves have become much more powerful, evolving from project management tools, with minimal additional reporting and data capture, to powerful suites that have functionality capable of supporting idea generation and development through benefits realization. They generally also have the ability to create and assign tasks, apply document templates to those tasks, and configure workflow that will move those documents and task assignments to different owners based on certain criteria.

This is where the benefit for risk management can really start to be seen, with the ability to automate some of the administration and tracking of risk management activities. Consider as an example the way a risk summary moves through the risk management process. The summary represents a single risk and is created as part of risk identification. It is then enhanced through risk analysis with the addition of more detailed information. This allows for the risk to be prioritized and for an appropriate management approach to be determined. The summary then gets an owner assigned and is updated as the management activities continue. A contingency plan will be created and referenced in the risk summary, and potentially the risk will trigger at some point. During all of this, the risk summary will be contained within risk management lists and risk management plans and once the risk management activities are completed, it will be archived.

This process can easily be accommodated within a PPM tool, and while different tools will handle the specifics in different ways, a typical approach will see a work item being created and the risk summary template being associated with that work item. The tools will allow for the definition of a workflow based on the type of work item that is being created—a risk in this case—and can then automate the movement of the risk based on different conditions being met. As a simple example, if the status of the work item is changed to *ready for analysis* and certain predefined fields in the template have been completed, then the tool can move the item into a work queue for the analysis to be carried out. Figure 24.1 shows how this simple step might work and how PPM software can help support that work.

Effectively the tool is automating:

- The creation of the basic project identification information—populating project identification fields in the template based on the project information in the PPM database, automatically assigning a risk identification number, and populating basic data like date of creation and person creating the risk.
- Validation of the mandatory fields—making sure that a risk is only sent for analysis when all of the data elements required for analysis have been provided. This can be a little dangerous as this is likely a binary check. If data is in a mandatory field, then the field is considered complete regardless of the quality or level of completeness of the data.
- Creation of entries in artifacts like risk lists and risk management plans. Again, this is based on the information submitted, so there won't be any consideration of the quality of the data.
- Monitoring of progress through alerts/reminders if work is taking too long or if updates aren't being made in a timely manner.
- Movement of the risk through the process to risk analysis along with information on age/time waiting. Potentially this can also include logic that will route the item to a specific risk analyst.
- Tracking and reporting of risk information through query and analysis functions that provide both high level summaries and

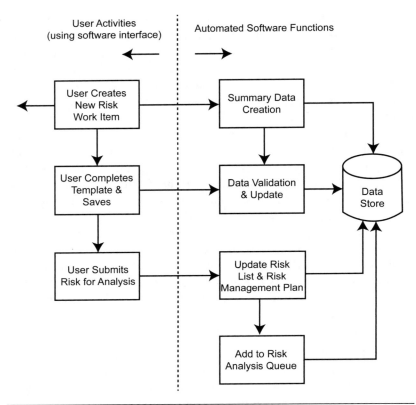

Figure 24.1 Risk summary creation and submission using PPM software

the ability to drill down to details of individual items, as well as the ability to filter reporting by multiple criteria (risk category, project, program, phase, vendor, severity, or age).

Clearly this is a useful support tool for the process—it reduces manual work and eliminates the possibility of human error in these automated areas. Of course, this automation can be extended to many of the hand-offs and templates that are used in the process. Furthermore, PPM tools contain powerful reporting and analysis engines that will make it easier to monitor what is happening within risk management activities, track metrics, and conduct analysis of everything from efficiency and effectiveness to the areas of maximum risk exposure and highest number of triggered risks.

However, it's not quite as simple as that. If you are planning to integrate your risk management process with a piece of enterprise software (and while PPM solutions are the obvious choice, other enterprise suites will likely have workflow and analysis capabilities), you need to recognize that the human element is being removed from the workflow that you are automating. That is helpful in reducing the chance of mistakes occurring due to people forgetting to update risk lists or risk management plans, but it also reduces the ability to deal with special circumstances. Automated workflows are based on the application of rules, and those rules will unfailingly be applied, even when we ideally want something different to occur—routing a risk summary to someone else for review before submission because we want some expert guidance, for example. In these scenarios, we end up having to work around the tool; inevitably, that involves inconsistent, undocumented steps and increased risk. Additionally, the rules ultimately rely on human intervention to be configured properly. If someone forgets to set their out-of-office status, then the rule that prevents tasks being assigned to someone who is out of office will never trigger!

Those aren't reasons to avoid using a PPM or similar tool to facilitate workflow, but they are reason to be careful about how heavily we rely on those tools. It's far better to require a manual confirmation step to trigger the automated processes than to end up fighting with the tool to get the outcome that you want. We also have to consider the extent to which the PPM software is utilized within the organization. We discussed earlier in the book how wide ranging the risk management process is—it involves the most senior executives and the most junior team members. If we are going to rely on a piece of software to automate elements of that process, then we need to be sure that all of those people have access to the system and the modules they will require. If that isn't the case, then the tool can still be used as a tool to support elements of the process—up to and including prioritization, for example. It can also provide an easily searchable archive of historic information.

Other Technology Considerations

Of course, there are other technology tools that can support process, and your organization likely already has a number of such tools in place to help with the management and administration of the other processes that exist. These should be leveraged wherever it makes sense to do so without compromising the efficiency or effectiveness of the process, but they shouldn't automatically be assumed appropriate. For example, if your organization uses a document management system with a number of metadata fields to help classify different process artifacts, then the metadata that is needed for operational or project level process may not be appropriate for organizational level processes. That doesn't mean that the tool isn't suitable, but it does mean that you likely need a separate workspace for your organizational processes where you can define appropriate metadata fields (and other characteristics).

The issue of data security also needs to be considered when tools like this are being used. In order to be fully effective, the tools need to be accessible to a large number of users. However, there will be some risks that are extremely sensitive, where visibility needs to be restricted. At times even entire programs will need to be conducted in secrecy, but they still need to be able to leverage the process. With today's technology, there is no reason why this level of restricted access can't be supported through a combination of configuration and user profile management, but care needs to be taken when the process is being integrated with the tool to ensure that the implementation can support all of the likely scenarios. This will always be a tradeoff between security and ease of access for a workforce that is likely spread across many locations, and potentially many countries, a workforce that won't always be working traditional hours from an office environment. Only you know where your organization places that tradeoff, but any technology tool you use to support your process needs to consider this balance.

We also need to consider that many of the people involved in risk management will have limited engagement with the process, both in terms of time and interest. A typical risk owner (if there is such a thing) will be assigned to a number of different tasks on the project or projects with which he or she is involved. Managing the owned risks is only a small part of their work. It's not a task likely to be top of the priority list because it's not directly associated with a task the owner's project manager is tracking. If we have rolled our processes out correctly, then risk owners will understand the importance of their risk management work. They will be committed to doing as good a job as possible, but they still need technology tools that support that.

If we present a simple web form that merely captures data and stores it but that is intuitive and easy to access and use, then risk owners will choose that every time to enter their updates. A sophisticated enterprise suite that requires a half day's training to even navigate to the right part of the application will not be embraced by end users, no matter how powerful the reporting and analysis is behind it. This is a reality that many organizations forget when investing in tools and is something you need to consider when deciding on the appropriate technology tools to use to support your process. Think of it as an exercise in risk management because that's exactly what it is—what are the risks (threats and opportunities) of choosing one particular tool over another.

Whatever technology tool or tools you decide to use, there needs to be consistency in the approach for that technology to be effective. A different tool may be used at different phases of the process; in fact that makes a lot of sense in many ways—risk management involves many different people and may require a different tool than risk analysis, which involves far fewer. Of course, we would need those tools to be able to communicate with one another to avoid the need for manual steps connecting them together. Where there is an issue is with the idea of different parts of the organization using different tools for the same elements of the process. If the process is being rolled out to all areas of the organization, then those areas have to use the same technology suite to support the process. If we allow different PMOs or departments to use different technology, then not only are we adding complexity (and by extension risk) to our ability to effectively manage the entire universe of portfolio risks, we are also creating an environment that will almost guarantee that the standardized process we have worked so hard to develop will begin to diversify almost immediately. The different technology platforms will inevitably drive differences in the way that the process is executed, which in turn will lead to the development of unofficial workarounds to overcome the challenges of a technology platform that doesn't integrate well with the process.

There are many other technology considerations that will be unique to your organization, but the single most important thing to remember when considering how to leverage technology to support your process is that the process always has to come first. You must never compromise the right process for the convenience of a tool—the risk just isn't worth it!

Index